NFORMATION LITERACY AND PLAGIARISM

(for Medical, Dental, Nursing Graduates and Allied Health Sciences)

W0080543

Ramesh Pandita

M.Lib.I.Sc, M.A (Sociology), M.B.A (IB) NET, M.Phil, Ph.D (Pursuing)
Sr. Assistant Librarian
Central Library
Baba Ghulam Shah Badshah University
Rajouri, Jammu and Kashmir

Shivendra Singh

B.Sc, M.Lib.I.Sc, Ph.D
Assistant Librarian

Baba Farid University of Health Sciences
Faridkot, Punjab

CBS
Dedicated to Education

CBS Publishers & Distributors Pvt Ltd

• New Delhi • Bengaluru • Chennai • Kochi • Mumbai
• Hyderabad • Kolkata • Nagpur • Patna • Vijayawada

INFORMATION LITERACY AND PLAGIARISM

ISBN: 978-93-86827-13-5

First Edition: 2018

Published by **Satish Kumar Jain** and produced by **Varun Jain** for

CBS Publishers and Distributors Pvt Ltd

4819/XI Prahlad Street, 24 Ansari Road, Daryaganj, New Delhi 110 002, India.
Ph: 23289259, 23266861, 23266867   Website: www.cbspd.com
Fax: 011-23243014
e-mail: delhi@cbspd.com; cbspubs@airtelmail.in.
Corporate Office: 204 FIE, Industrial Area, Patparganj, Delhi 110 092
Ph: 4934 4934   Fax: 4934 4935
e-mail: bhupesharora@cbspd.com

Branches

- **Bengaluru:** Seema House 2975, 17th Cross, K.R. Road,
  Banasankari 2nd Stage, Bengaluru 560 070, Karnataka
  Ph: +91-80-26771678/79   Fax: +91-80-26771680
  e-mail: bangalore@cbspd.com
- **Chennai:** No. 7, Subbaraya Street, Shenoy Nagar, Chennai 600 030, Tamil Nadu
  Ph: +91-44-42032115   Fax: +91-44-42032115
  e-mail: chennai@cbspd.com
- **Kochi:** Ashana House, 39/1904, AM Thomas Road, Valanjambalam, Eranakulam 682 018, Kochi, Kerala
  Ph: +91-484-4059061-62-64-65   Fax: +91-484-4059065
  e-mail: kochi@cbspd.com
- **Kolkata:** No. 6/B, Ground Floor, Rameswar Shaw Road, Kolkata-700014 (West Bengal), India
  Ph: +91-33-2289-1126, 2289-1127
  e-mail: kolkata@cbspd.com
- **Mumbai:** 83-C, Dr E Moses Road, Worli, Mumbai-400018, Maharashtra
  Ph: +91-22-24902340/41   Fax: +91-22-24902342
  e-mail: mumbai@cbspd.com

Representatives

- Hyderabad   +91-9885175004
- Vijaywada   +91-74069-04007
- Nagpur   +91-9021734563
- Mangalore   +91-9741432102
- Patna   +91-9334159340

Printed At : Goyal Offset Printers

Preface

The book, **"Information Literacy and Plagiarism"** is divided into 9 chapters. All the chapters are significantly important for understanding the need of being information literate and how the information illiteracy is becoming the reason for academic and research downfall across the world in general and India in particular. It has been constantly observed that researchers, despite being highly qualified fail to judge the most authentic, reliable and exact source of information. This leads them to become the victim of unreliable, fabricated and false information, which results in distortion of the facts thus, hampering the course of true research work. The book is targeted to benefit the research community across the nation.

In the prevailing IT environment, the internet is loaded with abundant information on all subjects, which results in coming across a number of sources of information that may divulge the information. The difficulty generally faced by the scholarly community is to choose the most appropriate source of information. In case of lack of awareness the task of searching the most authentic source of information may become increasingly difficult. The scholars simply choose any source and cite it, without getting into the authenticity and the reliability of the information. No doubt the information quoted may be 100% correct, but the source they are supposed to quote should be the one from where the information has actually originated. Therefore, under different chapters, an attempt has been made to draw awareness among the scholarly community about choosing the right sources of information and not to cite the information sources arbitrarily.

Plagiarism is equally a growing menace prevailing among the scholarly community. Needless to mention that plagiarism is about academic misconduct, whereby a good number of researchers, both seasoned and young label the work of others as their own. Plagiarism itself has different dimensions and therefore, researchers and academicians are confused about what actually amounts to plagiarism and what does not. With the introduction of plagiarism detection software, a gradual but steady awareness has come among the scholarly community that one can easily get hold any scholarly theft. Hence, one should no more indulge in any such academic misadventure, which can easily land a researcher in trouble or even their flourishing careers may be ruined. The fundamental awareness of what is plagiarism, what constitutes plagiarism and who will decide whether the content falls under the purview of plagiarism or not are some of the key concerns, which have been discussed in detail.

Software like, turnitin, iThenticate, etc. have been discussed along with their limitations.

The book deals with the different aspects of research writing, including reference management. Reference management is one of the most difficult, tiring and time-consuming aspects of the research activity. Accordingly, two chapters have been specifically introduced to help the budding researcher in the reference management, while writing research articles with the help of software like Mendeley and EndNote. The step-by-step guide will help researchers to easily download and use both the software, and can change the reference style anytime without wasting time. One chapter has been specifically added to address the legal issues involved with the software and the internet. The chapter discusses about various proprietary and non-proprietary software, open source software, free software, piracy concerns and copyright, etc. The last two chapters of the book deal with the aspects like, open access documents, digital documents and the resources and services of a typical academic library. We hope that you will enjoy reading the book and our efforts will prove faithful for you.

Ramesh Pandita
Shivendra Singh

Acknowledgements

First and foremost, our sincere gratitude to the Almighty *(Parampita Parmatama)*, for being so gracious and merciful and for blessing us with the everything from the commencement to the culmination of this book entitled "*Information Literacy and Plagiarism*", without which this book would not have seen the light of the day.

It is extremely difficult to acknowledge each person individually, who may have contributed, directly or indirectly, overtly or covertly, intentionally or unintentionally one or the other way during the course of writing this book. We extend our gratitude to all our near and dear ones for being there, without their support and cooperation it would not have possible for us to come up with such a masterpiece work. Our gratitude is due to our students, scholars, teachers and other fellow colleagues for being a constant source of inspiration. *Our gratitude and sincere thanks to Prof. Javed Mussarat, Hon'ble Vice Chancellor, Baba Ghulam Shah Badshah University for his support, cooperation and the encouragement.*

We are extremely grateful to **Mr Malik Mubashar Hassan, Head, Department of Information Technology Engineering (ITE), School of Engineering and Technology, Baba Ghulam Shah Badshah University**, Rajouri, Jammu & Kashmir for his significant contributions as a coauthor in framing Chapter-7 *Legal Issues in Software and Internet* of this book.

Our acknowledgments are also due to Prof. I A Hamal, Ex Vice-Chancellor, Baba Ghulam Shah Badshah University, Rajouri Jammu & Kashmir for his encouragement to come up with a book, which may address the issues of plagiarism because of which the society in general and the academic in particular is battling with and has become somewhat endemic to the modern day scholarly community. Our acknowledgements would be incomplete without mentioning the name of Prof. A.K Koul, Ex Dean, Academic Affairs, Baba Ghulam Shah Badshah University, who has been always a great guide, guru and a motivator, blessed with immaculate zeal, wisdom and foresight for being always there to share his expertize, whenever asked for.

We express our sincere thanks to Prof. Dr. Deepak J. Bhatti, Principal, Guru Gobind Singh Medical College and Hospital, Faridkot, Punjab for his support, cooperation and the encouragement extended.

Last but not least, acknowledgements are also due to Clarivate Analytics, Turnitin, URKUND, iThenticate, Microsoft Corporation and Elsevier for providing the training material to conduct the Mendeley training.

We would like to thank **Mr Satish Kumar Jain** (Chairman) and **Mr Varun Jain** (Managing Director), M/s CBS Publishers and Distributors Pvt Ltd for providing us the platform in bringing out book. We have no words to describe the role, efforts, inputs and initiatives undertaken by **Mr Bhupesh Arora** Vice President-Publishing and Marketing, PGMEE and Nursing Division for helping and motivating us.

We thank Dr Mrinalini Bakshi (Senior Content Developer and Editor) for her editorial support and Ms Nitasha Arora (Project Manager), Ms Neetu Jindal (Asst. Production Manager), Mr Nitish K Dubey (Senior Editor) and all the production team members Ms Tahira Praveen, Mr Ashutosh Pathak, Mr Prakash Gaur, Mr Chaman Lal, Mr Prabhat Ranjan, Mr Phool Kumar, Mr Bunty Kashyap, Ms Babita Verma, Mr Raju Sharma, Mr Manoj Chaudhary and Mr Vikram Chaudhary for devoting laborious hours in designing and typesetting of this book.

Contents

Chapter 1

Information Literacy: Need, Purpose and Importance

INTRODUCTION

Many a times, it has been observed that people use the term "Information Literacy" interchangeably with the "Literacy Information", which otherwise is altogether a different concept. In this chapter, attempt has been made to highlight the different aspects of information literacy (IL), as what information literacy is, its need and importance in day-to-day life of a common man and various other aspects, which we generally miss out for the want of information literacy. We have tried to highlight the key features of information literacy whereby it can be used as a tool to overcome the odds of life or one may confront in his/her research activity. In the prevailing information technology (IT) era of information and technology (IT) there is a far greater need to consume information in the same way and we consume our food and water and that too only after satisfying and ensuring ourselves from all aspects that the consumable is safe to eat. There is also a need to understand that the consumption of adulterated and contaminating information may have far-reaching consequences, the effect of which may even turn irreversible. Emphasis has also been laid on honing the skills toward traditional literacy, media literacy, computer literacy, digital literacy, network literacy and many more, as all these forms of literacy together constitute information literacy. Nowadays, in the information technology driven world, nobody can afford to be illiterate on any of these accounts.

WHAT IS LITERACY?

In the simplest terms, most of us recognize "*Literacy*" as the ability of an individual to be able to read and write, but this simple explanation does not actualize the true nature of literacy and a literate person. Literacy, in fact, has got with it different levels and degrees, which rest with each individual and varies considerably from person to person. The other level or degree of literacy can be termed as the ability to read, write and understand. Furthermore, literacy can be termed as the cognitive capacity of an individual, whereby he/she develops the power of judgment, analysis, synthesis, ability to take decisions and ability to recognize the nature of things in almost every activity of human life.

WHAT IS INFORMATION LITERACY?

Information literacy can be termed as the more extended aspect of the literacy, wherein people develop the capability to judge information by value. Information literacy is more about application of mind in assessing the credibility, reliability and authenticity of information about its use and application. Information literacy is more about having a set of abilities requiring individuals to "recognize—when information is needed and to have the ability to locate, evaluate and use effectively the needed information".

Each individual in his/her routine activity is put to mental or cognitive test frequently be it from purchasing routine grocery items to any other commodity. The foremost thing most of us do is to ensure the worthiness of all such commodities or products from all aspects and then go ahead with the purchase. An information literate person is supposedly having better knowledge, sound ability to judge things, clarity about the acts and actions of others and self. Accordingly, information literacy is about fine tuning of one's skills and abilities to imbibe all such qualities, whereby a person shows considerable improvement in his/her outlook toward the importance of judgment, ability in quantifying and qualifying things for their being along with their true worth and authority.

- Information literacy is not only about being able to locate relevant information, but also encompasses the ability of an individual to interpret, evaluate, and apply the retrieved relevant information.
- According to American Library Association (ALA) 1989, information literacy (IL) is about being definite about the problem.

COMPONENTS OF INFORMATION LITERACY

The various components associated with the information literacy include:

- Literacy
- Information
- Information literacy
- Information literacy (day-to-day activities)
- Information technology
- Information society
- Information-seeking behavior
- Need, purpose and importance
- Information haves and have nots
- Information filtering (Job markets)

COURSE OF INFORMATION

The authenticity and the credibility of information are the two important and inseparable aspects associated with the production, consumption and distribution of information. Information does not mean whatever we say, communicate or pass on to the next person should always be trusted. It has been observed that most of the times, people gather information from here and there or from various other known or unknown sources. They pass on such information to others without judging the authenticity, reliability and the credibility of the information. This behavior of humans to consume information without ascertaining authenticity has led the need to draw awareness among people toward information literacy, so as to judge whether the information provided is worth to be consumed or not.

Each piece of information moves through different stages. At each stage it is supposed to pass the test of reliability and authenticity, and this movement is generally seen in the below-mentioned order:

- Data (Collection of facts/figures from which conclusions may be drawn)
- Information (A message received and understood)

- Knowledge (Learning and reasoning)
- Wisdom (Ability to apply knowledge, experience, understanding or common sense and insight)

ALEXANDRIA PROCLAMATION OF 2005

It defines information literacy and lifelong learning as the "beacons of the information society, illuminating the courses to development, prosperity and freedom. Information literacy empowers people in all walks of life to seek, evaluate, use and create information effectively to achieve their personal, social, occupational and educational goals. It is a basic human rights in a digital world and promotes social inclusion in all nations".

INFORMATION AS A RESOURCE

Since time immemorial, the progressive movement of societies across the globe from one stage to another, is simply based on the accumulation of information and knowledge. It is this information and knowledge, which always played its part toward the growth and development of societies: Transition of human life from wanderers to settled societies, from hunters and food producers and covers all these phases of life.

The movement of societies was, is and will always be witnessed in the following forms:

- Humans accumulated information
- Information was converted into knowledge
- Information was propagated to further generation.
- It is the accumulation and exploitation of knowledge during the entire history of human civilization, which has led us to the present day of science and technology, what we often dub as an information society.

The moment societies would cease to follow the course of information and knowledge accumulation they would cease to grow. Information is one of the greatest resources in the society and forms the basis of every other resource available. There is no substitute for it. It also wont be perhaps inappropriate to say that every other resource available in the world or societies can be best exploited only when there is a sound information and knowledge base available with us about all such resources.

TYPES OF INFORMATION LITERACY

There is no end to the road of information literacy. Information literacy can be achieved or skills can be honed in any field and area and this enables a person to develop better understanding and knowledge about that subject or object. However, there are some common areas in which a progressive man always finds interest and hence, the need thereof arises to develop literacy in such areas. Some of the identified areas include:

- Traditional literacy
- Media literacy
- Technological literacy
  - Computer literacy
  - Digital literacy
  - Network literacy
- Sociocultural literacy

SOURCES OF INFORMATION

When we talk about the nature of information, its authenticity, reliability and various other components, there is an equal need to understand about the different sources of information. Accordingly, given below are some of the common sources and channels, having different characteristics, through which information is made available.

Formal versus Informal

Formal sources of information are those sources which are vested with authority, while the informal sources of information are not the authoritative sources of information.

Quantity versus Quality

Quality of information cannot be compromised with the quantity. However, quantification of information can be undertaken by means of the amount of information produced or the volumes of information produced.

Structured versus Unstructured

It is not always necessary that whatever piece of information we are be looking for will be available in the structured or desired form. Information which requires minimum efforts to put it into use can be termed as the structured piece of information, while the information which requires a lot of efforts for its structuration or which cannot be put into use in its available form is an unstructured piece of information.

Clutter versus Organized

Information which lies scattered across different sources and cannot be put to use unless same is organized in an any given state of need or requirement is known as cluttered information. Organized information on the other hand is more structured, which can be easily retrieved and readily accessible for use.

Primary versus Secondary

Primary sources of information are those sources in which first-hand information is published about any research activity or with regard to any other sphere of human activity. Periodicals like research journals, magazines, newspapers, bulletins, standards, serials, etc., are known as the primary sources of information. Secondary sources of information, on the other hand, are those sources, in which information is published in a more refined and more understandable manner. Secondary sources of information are mostly based on the information published in primary sources of information.

Popular versus Scholarly

Popular writings are generally based on the expertise of a person in any given field, wherein a person can write on the subject, based on his acquired knowledge from different sources of information. Popular writings are equally based on the individual understandings of a person about a subject or any other given topic. Scholarly writings are more based on verifiability. Scholars, who present some facts and figures about any given field support their findings and other writings by citing relevant sources of information, which they have consulted during their study.

Current versus Historical

Information pertaining to the latest developments in a society or in any given field can be termed as the current information. There may be various areas wherein no progress have been made or reflected for quite a while in all such cases, whatever information available lately is termed as the latest information on that subject or area. On the other hand, information which has a rich past of over hundreds or thousands of years can be termed as historical in nature. In historical information at many instances, we may come across those periods where there is no information available, historian generally termed it as black hole. Historians generally draw inferences about all such periods by looking at the developments which would have taken place prior and post black hole period.

Tacit versus Recorded

Tacit knowledge is that knowledge which rests within the mind of an individual and is more about the knowledge and expertise gained by an individual over a period of time in any given field. It is always difficult to record all the tacit knowledge of an individual, but it is always advisable to make maximum use of that tacit knowledge which rests in the minds of field and subject experts. Recorded knowledge on the other hand is that knowledge, which we properly record in a document and make it available for the coming generations use, mostly in the form of books and other recorded material.

Explicit versus Implicit

Information which is precisely and clearly communicated or readily observable leaving nothing to implication and withstands with the fact in accordance to the primary meaning of a term is known as explicit information. Implicit information on the other hand is termed as implied, though not directly expressed, but is inherent in the nature of something.

Online versus Offline

Both the online and offline sources of information are electronic in nature, but the fundamental difference between the two is that the online sources of information can be anytime easily accessed on the internet, as these documents or sources of information stand hosted on different web servers of the service providers across the world. Contrary to online sources of information, offline sources of information are also viewed, accessed on computer terminals, but are not available on the internet and are not hosted across the web servers.

Digital versus Print

Digital documents are simply the electronic documents available in different formats, be it text, audio, video, pictures, photographs or any other multimedia content, which can be accessed by using electronic gadgets like, computers, mobile phones, iPods, iPads, laptops, tablets or any other high-end electronic gadget. Print documents are those sources of information, which are available in printed format. One can go through the desired piece of information from the printed document and there is no need to use any additional gadget to access or use the information.

Born Digital versus Digitized

Of late, a growing trend has evolved among publishers across the globe to make the published matter available in both print and electronic formats. Although, both have their own advantages and disadvantages, but keeping in view of the increased use of electronic gadgets by the Internet-savvy people, publishers are somewhat forced to adopt the practices of hybrid publishing. With the growing use of electronic documents, the need is being felt to have the digital version of all those previously published documents, be it classic or any other form of published matter. Born digital are those documents or sources of information, which are being itself made available by the publishers to the public for use in digital form like CD, DVD and other online and offline sources of information. Digitization, on the other hand, is a process wherein the earlier published matter is converted into the digital form with the help of scanners and other digital techniques. All such digital documents are known as digital documents.

Open Access versus Subscribed

Open access has become the order of the day. Most of the research institutions across the world have started offering access to their published research articles free of cost. The scholarly publishers like Springer, Emerald, Taylor and Francis, Wiley, Oxford, and many other publishers seek the option from researchers about making their research papers available in the open access or closed access. Researchers, those who wish to make their research papers open access are generally asked for publication charges. Accordingly, the publishers provide access to their published material both in open access and subscribed form. Open access documents are those which are freely available to the information seekers online, while as subscribed sources of information are those for which one needs to pay for availing services or accessing those sources of information. Most of the academic and research institutions across the world subscribe to various sources of information for their users, mostly depending upon the information requirements of each individual institution.

INFORMATION FORMATS

Like different sources of information, there are also different formats of information. Some of the common formats in which information is made available include:

- Text
- Photographs, images, drawings, paintings
- Figures
- Verbal communication
- Audios/Visuals
- Multimedia, databases, datasets, data sheets, websites, etc.
- Physical gestures

NEED OF INFORMATION LITERACY

Although, there are numerous reasons, which may justify the need for having information literacy, but for our present discourse we have identified the following few areas, which warrant the need toward information literacy.

Unprecedented Growth in Information

During the first decade of the 21st century, the publishing industry has witnessed a sea change, wherein the industry has received a new facelift in its publishing techniques. Information technology, on the other hand, has empowered the publishing industry in its own way with the result the society at large witnessed the exponential growth in information production. In the scenario when there is such a large scale production of information all around, there is far greater need to develop the power of judgment toward assessing the originality and authenticity of the information, as what information is worth to be trusted and what not. Information literacy empowers an individual in its own way, whereby one becomes enough able minded to access, retrieve and use a particular piece of information.

To judge information by value

Cognitive capacity varies from an individual to individual. No two minds can think and understand a concept in the same way by using same piece of information. This can be interpreted in the same way wherein we can say that two individuals do not enjoy the same social standing, despite having equal opportunity to resources and services. Individuals who are able to make better use of resources and services, undoubtedly they are better placed. Thus, the individual's ability to judge the value of each piece of information plays a very crucial role in his/her social standing. Information literacy on the other hand is a tool or a technique, which enhances these abilities in an individual, whereby one finds self as an empowered individual, having better abilities toward decision-making.

For Personal Empowerment

Personal empowerment is more about developing self-confidence and ability to make decisions, which are result oriented. Information literacy enables a person to develop his/her cognitive capacities, whereby an individual finds himself in a position to contribute both to his personal and professional life with a greater degree of satisfaction.

For Economic Development

There is no denial of the fact that information has got economic attributes and the more we are able to use and apply information on any chosen economic activity, the greater are the chances that our efforts may bear a good fruit. Information literacy can be correlated with developing skills in the field of economics, which enables a person to make best in the economic activities.

Helps to Navigate in Digital Environment

The amount of information, which is available on the Internet across millions of web servers, it becomes difficult for a information seeker to come up with a particular piece of information as the most reliable, authoritative and exact. The greatest difficulty with the sources of information available on the net is that most of the information available is duplicate and still more, varieties of explanations stand available on a single subject or topic which makes it further difficult for a user to refine his/her search. Information literacy or the technological use of literacy is more about zeroing ones choice on those sources of information, which are reliable, authentic and authoritative, hence the navigation across web sources becomes easier and result oriented.

Information Seeking and Use

During the last decade, information-seeking behavior of the individuals has changed a great deal. People, who otherwise used to rely heavily on the printed sources of information for their information requirements, have gradually started switching over to the online sources of information. Previously people who used to publish their research papers through conventional means viz. print has switched over to online publishing methods. However, there are people who oscillate between both the online and printed sources of information. In a nutshell, there is not a particular source of information, which can be termed as the most sought after, but surely the demand for none of the sources seems to have decreased. Information literacy enables us to find the differences between the different sources of information and also helps us to develop the competency whereby one can choose the best sources of information to quench his/her information thirst.

Learning to Search the Web

Information available across web servers is more in an unorganized and unstructured form, therefore, the inability of an individual toward making effective use of web-resources add to the problems and makes it more difficult for the information seeker to decide as what to trust and what not. Seeking a relevant piece of information from web resources is both a matter of skill and cognitive ability of an individual. Information Literacy is one such vital exercise, which enables an individual to develop both the skill and the ability to locate the desired piece of information from the huge clutter spread across the web.

SOME OTHER FACTORS

Apart from above-mentioned factors there are various other reasons, which advocate the need for information literacy.

- Information literacy helps one to define one's information needs, access, process and use the retrieved information strategically for attaining one's personal, professional and educational goals.
- Individual perceptions of people come into play to judge the authenticity and reliability of information they are being provided or supplied with
- Each single activity of every individual is bound to generate one or the other sort of information.
- Access to different sources of information for effective and efficient information retrieval
- Evaluate information and its sources critically and incorporate selected information into knowledge base and value system
- Use information effectively to accomplish a specific purpose individually or as a member of a group
- To understand the economic, legal and social issues related to use of information and access the information ethically and legally.

INFORMATION FOR ALL PROGRAMME (UNESCO)

The Information for All Programme (IFAP) was initiated by the UNESCO in the year 2000 on an intergovernmental basis with the view to create equitable societies by providing better access to information across all societies. Technological advancement is being seen as one good reason, which apart from revolutionizing the use of information has also increased its widespread access. Information is being seen as key and central to the each and every kind of development activity taking place all across the globe and if, a society is to be developed then there is a need to extend equal opportunities to all societies to access the sources of information so that they can transform their life by making good use of information.

The UNESCO has identified the following areas wherein the IFAP can be effectively channelized.

- Train the trainer
- Development of comprehensive information literacy programs at all educational levels
- Teacher and librarian training
- Create awareness in governmental and civil society institutions
- Research and its dissemination networks
- International cooperation in actions with the use of indicators and observatories.

WHAT IS AUTHENTIC INFORMATION?

Following are the very few attributes, which corroborate the authenticity of information from various angles.

- Availability or provision of the right kind of information
- Free from false or misrepresentation of facts
- Trustworthy
- Supported by evidence
- Vested with authority
- Having a legal validity

- Verifiable to establish its genuineness
- Credible, competent and reliable

PURPOSES OF INFORMATION LITERACY

Information literacy runs almost on parallel lines to that of mission raised across different nations toward increasing the literacy rate among wider cross sections of society. People have been able to understand the need, purpose and importance of real education and how actually education helps an individual, a family, a section, a group, a community to grow, develop and ultimately become a part of national mainstream and the same holds true about information literacy. Information literacy in a way can be termed as the application aspect of general literacy, wherein an individual finds himself or herself in a position to make best use of his/her literary attainments.

- Developing professional competence
- Value and use of information
- Skill development
- Bridging knowledge gap
- Flow of information from haves to have nots
- Information economy.

BENEFITS OF BEING INFORMATION LITERATE

Information Literacy helps us in a variety of ways to locate the exact and the required piece of information. Some common benefits of being information literate are:

- Considering appropriate sources of information
- Using information to learn
- Making use of information in a variety of ways
- Developing information competency to access, evaluate, organize, exploit, create or even disseminate information
- Developing skills to extract most authentic, reliable and exact information.

WHY AND HOW TO PROVIDE INFORMATION LITERACY?

Promotion has become an integral part of a modern day life. An industrialist promotes his/her products to survive in the market and to attract people toward his products. By marketing products and services, the industrialist makes people aware of his product by showing similarity and difference with that of other similar products available in the market from the rival groups. This way, people will be left with a choice to go for a particular product. Information literacy on the other hand is more related with a social cause and concern. There is every need that people should get benefitted from every such activity or action which is aimed for their welfare and betterment and the information illiteracy can act as an impediment in their way. All this gives us a better reason to promote information literacy among people, as this will help people to survive and sustain the ever-increasing pressure to differentiate between the original and fake, authentic and pirated, reliable and unreliable and many more other concerns, which have become the reason to promote information literacy among people.

In schools, colleges, universities and other academic institutions, the information literacy can be promoted among students by working and adopting the following practices.

- Teaching as a course in academics
- Part of curriculum
- Teacher education
- User education/orientation
- Faculty development Programs
- Experiencing information literacy (learning),
- Reflection on experience (being aware of learning) and
- Application of experience to novel contexts (transfer of learning).

INFORMATION INSIGHTS

When we talk about information insights, it means we are trying to draw the attention of the audience or the readers toward those aspects of information, which, although people may be knowing superficially, but the informative insights help one to develop clear and deep perception about the use and abuse of information. Informative insight is more about developing the feeling of clear understanding, grasping the inner and the exact nature of information. Informative insight helps to develop understanding about handling the complex information. It also helps to

- Understand ethical, legal and socioeconomic issues surrounding information and information technology
- Following laws, regulations, institutional policies and etiquette related to the access and use of information resources
- Acknowledging the use of information sources in communicating the product or performance.

JOB MARKET FOR IL PROFESSIONALS

Job markets for information professionals are—searching, retrieving, filtering, structuring and ultimately its provision for consumption has immense potential. Library and information professionals are being employed in most of the service sectors and their foremost role is to identify and filter the information for use. Besides, given the amount of information available across different sources, be they print, online or offline, it becomes literally difficult for a researcher to locate the exact piece of information. As locating the information from several sources is not only the tiresome, but also a time-consuming activity, which often delays the research results beyond the expected time. In order to overcome this problem, need is being felt to handover the job of locating authentic information to information professionals who are well versed in this activity. This will help researchers in a great deal in devoting their maximum time toward the research activity rather spending time in locating the information.

Need of information specialists is also being felt across, industries, corporate houses and other private, public and government sector organizations. These organizations are also hiring the services of information professionals toward locating the authentic information for further use by business establishments. Morrison and Stein (1999), in their study while viewing their concerns about the information explosion, observed that more than 20% of all the future jobs will be involved in locating and checking the authenticity of the information to be consumed, by assessing the data and thereafter provide the same for consumption.

INFORMATION LITERACY AND RESEARCH PROBLEM IDENTIFICATION

Information literacy plays a very significant part in identifying the research gaps and in formulating the research problems. An information literate person always finds himself/herself in better position wherein he/she can identify the information needs to fill any existing information or research gap. Some of the key activities and areas, which an information literate person can handle more efficiently include.

- To identify a research topic or other information needs
- Develops a thesis statement and formulates questions based on the information need
- Identify information sources to increase familiarity with the topic
- Describing information need
- Recognizing the need to correlate existing information with potential research area
- The ability to know the means of information production, organization and dissemination.

IL AND INVESTIGATION METHODS

- Lab experiment
- Simulation
- Field work, etc.
- Information literacy and search strategy

IL HELPS IN CONSTRUCTING EFFECTIVE SEARCH STRATEGY

Searching the most authentic and reliable information across different resources is equally a difficult task. Information seeker has to understand that in order to make their search easier, smoother, better and above all, result oriented, information literacy is going to be a way out. Information literate people are always able to frame a better search strategy. Unless an information seeker is not having the basic understanding of some of the fundamental search techniques, he/she cannot retrieve the desired information in a timely manner. An information literate person is generally able to

- Identify key words and synonymous terms
- Use of controlled vocabulary (Topic specific)
- Boolean Search (AND OR NOT)
- Truncation ($) or (?)
- Search engines
- Indexing and abstracting services, etc.
  - Indicative and informative abstract

IL TO SEARCH LIBRARY RESOURCES

Many a times, it has been observed that library members are not able to locate a particular document in huge collection of the library. This again can be termed as a problem more associated with lack of application of mind and wisdom in getting the needful done. Even there may be library users who may not be aware of this fact that documents on the library stacks are arranged in a scientific manner and not the arbitrarily, what they generally presume. So there is need to train the library members about the use of library resources. Given below are some of the common tools and techniques used by the library professionals to organize the library resources in more structured and scientific manner.

Classification Schemes

There are different classification schemes available and used by library professionals in arranging the library documents in order. These classification schemes are generally enumerative in nature, mostly based on Indo-Arabic numerals and a few others having English alphabets. Classifying documents are more a scientific arrangement, wherein all the related documents come together, which facilitates their easy retrieval. Classification is more about translation of bibliographic details into ordinal numbers, which includes, title, subject, etc. The library clientele are supposed to have the necessary knowledge about the classification numbers and the type of arrangement, which is carried out in libraries for documents. An information literate person can easily locate and retrieve the documents from the library resources. Some simple and basic technicalities involved with the classification schemes are discussed below.

Book Numbers

Book number is a tool used for translating the name of the author into ordinal numbers. Book number simply lends an extension to class number and is very useful in locating a particular document in a library by a particular author. There are different methods of assigning the book number to a document, depending upon the individual preference of each individual library. Even in some of the libraries, the first three or four letters of the author are used as book number. However, the cutter book number table is generally preferred by most of the libraries in assigning the book number.

Call Numbers

Call number helps in fixing and locating the position of the document in the library. The call number is a combination of the class number, the book number and the collection number of a document, which facilitates the retrieval of a document from the library collection. Assigning a class number, book number and collection number to a library document reflects that the minimum necessary technicalities required for fixing the location of a document has been undertaken, so as to facilitate its easy retrieval.

Online Public Access Catalogue

Online public access catalogue (OPAC) is the technological offing to the library services, wherein it facilitates in identifying a particular document in the library collection and also reflects its exact location, its availability in the library and in case the document is already on loan, then it's possible date of return along with the reservations, if any of the document, etc. The modern day catalogue is more helpful and easier to use than the conventional cataloguing practice. In the conventional cataloguing, it was always time consuming to locate a particular document, therefore, it was difficult to ascertain whether the document is on the shelf or not, and to take the repeated searches was equally an uphill task. Even to trace the document from the borrowers list was never an easy practice. On the contrary, with the help of OPAC one can take repeated searches for his/her document within seconds and can get the document issued himself/herself at the self-check counters without any human help. Library members or library clientele having awareness about the OPAC facility can make a good use of library resources and services.

Human Help

Indeed, technology has helped us in a great deal in overcoming the difficulty of locating the document in huge collection of a conventional library system, but all that has not undermined the role of human help in assisting and locating a document from the stacks. It is the primary role of the library professionals to arrange the resources on the stacks in the most helpful manner, so that library clientele may not find it difficult to retrieve the desired documents. Also, we can make a good use of technology in each service and activity area of the library, but it is always the library professional who is supposed to carry out all such activities and render services in his/her personal supervision. So the role of human help can neither be undermined nor can be taken away in rendering efficient library services. Library member having minimum basic knowledge about the library system can retrieve a document of his/her choice from the library without any human help.

Interlibrary Loan Facility

Interlibrary loan (ILL) is one of the extended library services, wherein, if a document is not available in a particular library, the same is arranged from any other library on loan basis. This kind of service is generally provided when the two libraries or institutions sign a memorandum of understanding about providing such services. Provision of ILL services at local level is very helpful, as by entering into an agreement to provide ILL services, the local libraries can avoid raising the duplicate collection. Libraries can add those resources which may not be available at other local libraries with which one having ILL understanding. Library members having awareness about the ILL facilities are surely going to be the ones who can make the maximum use of the library services.

Document Delivery Kiosk

Nowadays, in the technology-driven world, even the document delivery is itself undertaken by the library members. The library members simply need to visit a library document delivery kiosk, which generally remains available at the exits of all the large libraries having technology-enabled services, can easily get their document issued without help of any third party. The document delivery kiosks are equally used for returning the documents in the same way the documents are generally issued. The more a library member is aware of all such technological offing's the more one can make use of such facilities.

Online Sources

Subscribed

There is a growing demand among researchers and academia toward the subscription of online sources of information. Although, nowadays libraries are subscribing both the conventional viz. printed sources of information, as well as online electronic sources of information. People who still reel under the conventional methods of accessing the information are somewhat deprived of the information published in the online sources. There is far greater need for the user community to be aware of all such sources of information of which otherwise they can make a good use of. There are people, who suffer with technophobe, as such, are unable to make a good use of all such online resources and services. There is always a need to subscribe the sources, which the institution wants to. Without subscribing the sources users will not be able to access those resources and services.

Open Access

The other forms of the online sources of information are the open access sources. The open access sources of information are those sources which are made freely available to the user community all across the globe. There are millions of quality documents, which are being made these days available to the user community free of cost. People not having awareness about all these open access resources are not able make use of all such resources. The open access documents include books, newspapers, journals, magazines, standards, etc.

Accordingly, the library members can access information on a variety of subjects by availing various other services, which may include.

- Institutional research cell
- Subject experts/practitioners
- Community resources, etc.

13

- Indexing services
- Abstracting services
- CAS services
- Selective dissemination of information (SDI) services
- Content management services
- New arrival services
- Bibliographic service and many more

INFORMATION LITERACY AND INFORMATION PRODUCTION

As we say information literacy is about fine tuning of one's ability to judge and look for the most authentic and reliable information, accordingly information literacy proves helpful toward the production of new information. Information production by no means should be treated as an industry, wherein purpose or aim is only to produce new information, but is more about producing the need based, authentic and the reliable information. The consumers of such information by no means should feel cheated for being provided or supplied with the contaminated, false and unreliable information.

Some of the common forms of creating new primary information include.

- Experiments
- Survey
- Interview
- Letters, etc.

Production, creation or generation of new information is not an end itself, there is always far greater need that newly created information should be presented in such a way that its users must find it easier to consume, apply and reap dividends. An information literate person has the ability whereby he/she finds himself or herself always in a better position to represent newly generated information in most convenient, easy to understand and appropriate format.

INFORMATION LITERACY AND LITERATURE REVIEW

Literature review is one of the essential and foremost components involved in the creation of a new piece of information or knowledge. It is always the existing pool of knowledge in any given subject or field, which forms the basis for any new knowledge about to be created. Reviewing the existing literature is a very specialized task and to correlate the existing knowledge with the new findings requires the subject specialization along with ability to put the findings in a more cohesive manner. While reviewing the literature information literacy helps an individual to develop the following skills:

- Competency to select main idea
- Correlating and restating the text as per his/her understanding of ones work
- Verbatim competency
- Examining and comparing information collected from various sources
  - Accuracy
  - Authority
  - Relevance
  - Reliability
  - Validity

Apart from above, while undertaking any research activity, a researcher has to take into consideration the following aspects associated with the production of a new piece of information.

- Synthesize ideas to construct new concepts
- Making use of technological applications for data analysis etc.
- Finding limitations in the information gathered
- Understanding, interpreting and validating information
- Ability to draw satisfaction from the information retrieved
- Ability to integrate new and prior information
- Drawing conclusion on the basis of information gathered
- Determine the extent of information needed
- Access the needed information effectively and efficiently
- Evaluate information and its sources critically
- Incorporate selected information into one's knowledge base
- Use information effectively to accomplish a specific purpose
- Understand the economic, legal and social issues surrounding the use of information, access and use information ethically and legally.

Information Literacy and Ethical Issues

Ethics form the very basic essence of any activity or action an individual undertakes or executes, and it these ethical issues, which people mostly confront for the lack of awareness , which otherwise can be easily overcome with the help of information literacy. It is always important for an individual to be bound by those ethics, which warrant an activity or an action be undertaken or executed on certain desired lines. Professional ethics in a way help to regulate the things within certain confines, which ensure their credibility, authority and authenticity. Information ethics help a person in accessing, retrieving, collection, and exploitation of information. Information literacy helps an individual in assessing the use of information in following areas.

- Understanding the ethical, legal and Socioeconomic issues concerning information and information technology
- Intellectual property rights
- Copyright
- Censorship and freedom of expression
- Plagiarism, self-plagiarism
- Upholding, institutional policies
- Attributions, citations, permission letters to use copyrighted material, etc.

PUBLISHING RESEARCH RESULTS

Publishing research results constitute an important and integral component of the information production and its dissemination. Any research activity undertaken by all means remains incomplete, if the research results of the same are not made public for the ultimate beneficiary or the end user. Each piece of information we produce on a day-to-day basis is bound to be consumed by good section of a society.

Publishing information or research results may by and large sound an easy job, but there is every need to understand this basic fact that it is the purpose which should be ultimately fulfilled and if the purpose is not fulfilled then most of the efforts go in vain. The aim should be that maximum people should get benefitted by any new piece of information or knowledge, and if we fail to make that information to reach to maximum users, then we surely somewhere misfit in publishing our research results. Accordingly, a researcher or a writer after careful consideration has to choose the most appropriate medium, which may prove most handy in disseminating the information to the largest population at least cost. A researcher has to see to it that what sources of information are there which are being mostly viewed by the end users in any given field. This will help a researcher to maximize than outreach. Here, it again becomes the reason wherein a researcher or a writer needs to be aware of all such sources of information, which can help him in fulfilling his purpose of publishing a particular piece of information for broader and deeper reach.

There are a range of services wherein a researcher can easily locate or identify the most appropriate source of information with wider reach to publish his/her research results. Some of the resources and services available and easily accessible to academia are discussed below.

Indexing Services

Indexing is simply a technique wherein a information seeker is able to locate the desired piece of information in any given document without wasting any time. Indexing services include those services, wherein an information seeker is able to seek the bibliographical details of a particular document in which a particular piece of information stand published. As per the British Indexing Standard (BS 3700: 1988) an indexing service is a systematic arrangement of different contents to locate a particular piece of information in any given document. Generally indexes are given at the end of each book, journal or any other source of information to facilitate the easy location of published content. These indexes are generally given in the alphabetical order along with page numbers, etc. Accordingly, indexing services associated with journals or other periodicals deal with the aspects like the articles, commentaries, statutes, official reports, etc. published in any given field. Similarly, we have indexes dealing with legislation, jurisprudence, etc.

Some of the common indexing services available among the academia and research community of the world include

- Indian Science Citation Index (ISCI)
- Indian Health Science Citation Index (IHSCI)
- Indian Agriculture Citation Index (IACI)
- Indian Social Science and Humanities Citation Index   (ISSHCI)
- Indian Journals Citation Report (IJCR)
- Indian Science and Technology Abstracts (ISTA)
- Directory of Indian RandD Journals (DoIJ)
- NUCSSI (Scientific Serials) (NISCAIR)

Thomson Reuters, Elsevier's, SCOPUS and a large number of discipline-specific indexing services are there

Abstracting Services

Abstract is a nutshell summary of the whole document, wherein the focus is mostly laid on to reflect the key aspects of the entire document, results and other essential components. Abstracts are formal documents, generally issued by the original authors about their own work, which helps in giving an insight to information seeker about the whole document, the key components, discussion, findings and other summary. As per the Encyclopedia Britannica (1964) an abstract is a condensed summary of the essential facts of theories and opinions presented in an article or a book. An abstract can be about any academic or nonacademic piece of writing, generally written anything between 100 to 300 words or so. Most of the abstracts are generally about research articles, thesis, reviews, conferences or any other type of document. An abstract simply gives a central idea about the whole document to the reader and it is the reader who has to decide after going through the abstract that whether he/she should go through the whole document or not. These kind of sources of information helps a great deal in saving the time of the users, especially when the researchers are already reeling under the pressure of overabundance production and availability of information on any given topic.

Abstracts are generally of two types, one is informative and the other is indicative. Informative abstract is that abstract, which may reflect the findings of a document to such an extent, wherein the seeker of information may need not to consult the original or the whole document for his information needs and the same may get fulfilled by going through the abstract only. On the other hand, indicative abstract is that abstract, which only gives an idea about the information produced or published in a particular document. It depends upon the information seeker whether to go through the where document or not by going through the abstract only.

Some of the commonly offered abstracting services across academia include,

- Indian Science Abstracts
- Medicinal and Aromatic Plants Abstracts
- LISA
- Chemical Abstracts
- Mathematical Reviews
- Medline

Impact Factor

Impact factor is one of the quality parameters used across the academia to judge the quality of a research article or the research journal. The impact factor is calculated on the basis of citations received by a research article or a research journal during a particular period of time. The concept of citation count was first suggested by Gross and Gross in the year 1927 and it was Garfield, who in 1957 was instrumental in devising the formula to compute the impact factor. Impact factor can be computed for any given period of time, but it is always desirable to compute the current impact factor of any research article or a research journal. As per the Web of Knowledge, Journal Impact Factor (JIF) is "the average number of times articles from the journal published in the past two years have been cited in the Journal Citation Report (JCR) year."

Researchers and academicians across the world prefer to publish their research results in those journals or periodicals which have a high impact factor. It is an established fact that research journals having high impact factor are highly popular among the scholarly community, both to access and to publish their research results. In fact Impact factor of a research journal acts as a parameter to publish the research results in any given research journal.

H-index

The concept of H-index was given by Jorge E Hirsch a UCSD (University of California, San Diego) physicist in 2005. Like impact factor, H-index is also used as a quality parameter among researchers to quantify the quality of a research publication. The concept runs parallel to that of impact factor as it too is calculated on the basis of number of citations received by a researcher, a research publication or a research journal during a particular period. As per the web of science, H-index is "an index that quantifies both the actual scientific productivity and the apparent scientific impact of a scientist.

Researchers and other academicians also use the H-index as a parameter to judge the quality of a research journal to publish their research results. The greater the number of citations a journal receives, higher will be its impact factor and more popular it will be among the scholarly community. Researchers intending to publish their research results can take the H-index of any research journal as a good reason to publish his/her results.

17

BIBLIOGRAPHY

1. Amaracian Library Association (n.d). Information Literacy competency standards for higher education. [online] Available from www.ala.org/acrl/sites/ala.org.acrl/files/content/standards/standards.pdf [AccessedJuly 14, 2016]

2. Andretta S. Information literacy: A practitioner's guide. UK: Chandos Publishing; 2005.

3. Association AL. Information literacy competency standards for higher education. 2000;1:63-7.

4. Beacons of the Information Society: The Alexandria Proclamation on Information Literacy and Lifelong Learning. [online] Available from www.ifla.org/publications/beacons-of-the-information-society- the-alexandria-proclamation-on-information-literacy. [Accessed on December 11, 2017]

5. Behrens SJ. A conceptual analysis and historical overview of information literacy. College & research libraries. 1994;55(4):309-22.

6. Breivik PS, Senn JA. Information Literacy: Educating Children for the 21st Century: ERIC. 1994;198.

7. British Standards Institution. 1988. British standard recommendations for preparing indexes to books, periodicals and other documents. 2nd rev ed. London: British Standards Institution; BS 3700 1988.

8. Bruce BC. Literacy in the Information Age: Inquiries into Meaning Making with New Technologies. ERIC;2003.

9. Bruce CS. Workplace experiences of information literacy. International Journal of Information Management. 1999;19(1):33-47.

10. Bruce CS, Candy PC, Klaus H. Information literacy around the world: Advances in programs and research, volume 1. Wagga Wagga, New South Wales: Centre for Information Studies, Charles Sturt University; 2000.

11. Darwin C. Encyclopedia Britannica. Chicago: William Bento Publishers; 1964;7:83-4.

12. Doak CC, Doak LG, Root JH. Teaching patients with low literacy skills. AJN. 1996;96(12):16M.

13. Doyle CS. Information Literacy in an Information Society: A Concept for the Information Age. Darby: Diane Publishing; 1994.

14. Eisenberg MB, Lowe CA, Spitzer KL. Information Literacy: Essential Skills for the Information Age. ERIC; 2004.

15. Garfield E. Citation indexes to science: a new dimension in documentation through association of ideas. Science. 1955;122(3159):108.

16. Grafstein A. A discipline-based approach to information literacy. The Journal of Academic Librarianship. 2002;28(4):197-204.

17. Grassian ES, Kaplowitz JR. Information Literacy Instruction: Theory and Practice. Information Literacy Sourcebooks. ERIC; 2001.

18. Gross PL, Gross EM. College libraries and chemical education. Science. 1927;66:385-9.

19. Herring JE. Improving Students' Web Use and Information Literacy. UK: Facet Publishing; 2010.

20. Hirsch JE. An index to quantify an individual's scientific research output. Proceedings of the National academy of Sciences of the United States of America. 2005;102(46):16569-72.

21. Iannuzzi PA, Mangrum I, Charles T, Strichart SS. Teaching information literacy skills. Boston, MA: Allyn & Bacon;1999. pp 1-200.

22. Journal citation reports: Impact factor (2015). Web of knowledge. [online] Available from www.admin-apps. webofknowledge.com/JCR/help/h_impfact.htm. [Accessed December 11, 2017]

23. Kafai Y, Bates MJ. Internet web-Searching instruction in the elementary classroom: Building a foundation for information literacy. School Library Media Quarterly. 1997;25(2):103-11.

24. Kubey RW. Media literacy in the information age: current perspectives, Vol. 6. Livingston campus of Rutgers University: Transaction Publishers. 1997.

25. Lankshear C, Knobel M. Digital literacies: Concepts, policies and practices, Vol. 30. Switzerland: Peter Lang Publishing; 2008.

26. Loertscher DV, Woolls B. Information literacy: A review of the research. Castle Rock CO, USA: Hi Willow Research and Publishing; 2002.

27. Lupton MJ. The learning connection. Information literacy and the student experience. Australia: Auslib Press Pty Ltd; 2004.

28. Morrison JL, Stein LL. Assuring integrity of information utility in cyber-learning formats. Reference services review. 1999;27(4):317-26.

29. Pappas CC. The information book genre: Its role in integrated science literacy research and practice. Reading Research Quarterly. 2006;41(2): 226-50.

30. Posner MI, Rothbart M K. Educating the human brain. APA. 2007. p 263

31. UNESCO. Information for all programme. [online] Available from https://en.unesco.org/programme/ifap" [Accessed July 18, 2016].

Chapter 2

Issue of Plagiarism in Higher Education

INTRODUCTION

The chapter discusses about the prevailing academic misconduct among the scholarly community which is more often correlated with the moral education, which academia generally ignore for various reasons, especially in the prevailing IT environment, wherein technology has become one of the most propelling forces. The concerns have been raised toward the growing menace of plagiarism among the scholarly community, which of late, has started coming to fore in numbers, due to the presence of plagiarism detection software. There are many evidences wherein prominent people have been found guilty of plagiarism, ultimately brought disrepute and end of the glorifying careers. An attempt has been made to give a very strong and clear message to the budding scholars that at this stage they may pay a deaf ear to the concerns raised about the growing menace of plagiarism, but five or ten years down the lane, any act of plagiarism detection can easily ruin their flourishing careers. Research supervisors being coordinator of the research work have a very important role in checking the practice of plagiarism, thus can easily regulate the research ethics among scholars.

Plagiarism in research is one of the important areas of concern, with which academia all across the world is directly associated. Plagiarism has been derived from the Latin word plagiary, which means to abduct somebody or to enslave. In academic or scientific writing, it means stealing somebody else's words or sentences and passing them off as one's own. As per Oxford Dictionary, plagiarism "the practice of taking someone else's work or ideas and passing them off as one's own", as per Cambridge Dictionary "to use another person's ideas or work and pretend that it is your own". "To steal and pass on (the ideas of words of another) as one's own or to commit literary theft" by The Merriam—Webster Dictionary. This practice is considered unethical and infringement of copyright. There is always far greater need to look at things beyond what they appear to be. The following points are discussed to know the plagiarism in higher education.

- Different forms of academic misconduct.
- Why do students/researchers do it?
- How can it be avoided?
- The role of library professionals?
- What are the different plagiarism detecting services available?

OCCURRENCE OF PLAGIARISM

Although, the prevalence of plagiarism dates back to 1970, when people were caught having stolen the ideas of others and presented those ideas as their own in their work. This was perhaps the period when the writers, perhaps felt that their theft may not get noticed. As societies moved forward, information and technology (IT) became an integral part of the modern day life and so did societies start relying on IT which IT in turn proved instrumental in detecting the acts of plagiarism. The acts of plagiarism reached to the level of copying and pasting stage with no extra effort. All this increased the instance of plagiarism to a much greater degree, as more and more people started getting involved in this undesirable practice. iThenticate (2012) in their study, involved more than 400 researchers, authors and editors, revealed that 1 in 3 editors regularly encountered plagiarism. 95% of editors and 84% of researchers reported that they occasionally or regularly encountered instances of plagiarism. 60% of those survey believed that plagiarism was on the increase. The same study found that the actions to prevent or detect or expose plagiarism were often not taken.

THE ISSUE

Generally, it is being observed that people involve in the acts of plagiarism are not actually aware about the gravity of the problem associated with plagiarism and even the consequences, which their actions may bore if caught in the act. The students, scholars and other researchers know it very well that any of their act involving academic fraud is totally unacceptable, and above all against the ethics of the academic and scholarly publishing.

NEED TO KNOW ABOUT PLAGIARISMS

India is one of the fastest growing economies in the world and the whole world is looking at India as a potential global leader, the largest consumer market in the world, the fastest growing industrial country with ability to produce quality products at minimum cost and many more reasons. Among so many emerging areas, higher education is one of the potential areas, which has started opening its vistas across the globe. Students from different countries across the world prefer India as a potential destination to fulfill their ambitions of higher education.

Institutions of higher education have to make sure that they should never compromise with the quality of education and research. Poor research is bound to bring disrepute to the higher education system of the country in general and the institution in particular. Acts of plagiarism in research are those few gray areas, which deserve to be given greater heed. Accordingly, the governing bodies of higher education of late have started offering plagiarism detection services to universities and colleges of higher education and research, so that academia in general and the scholars in particular be sensitized about the menace of plagiarism prevailing all over the world. By extending the services of plagiarism detection tools, aim is to draw awareness across the research community of the country to follow research ethics and not to indulge in undesirable practices.

The researchers all over the world have to make sure that their research is free from plagiarism. It becomes imperative for researchers to ensure ethical research for following reasons.

- UGC, India has through UGC Notification 2009, made it mandatory for the universities to check M Phil and PhD theses submitted to them through plagiarism detection web-based tools like Urkund, Turnitin and iThenticate. At present all UGC-funded universities have access to plagiarism detection software Urkund through INFLIBNET.

- The various international publishers of scholarly journals have incorporated the use of anti-plagiarism software tools into their editorial and publication process. It means that the editors check the manuscripts for similarity by scanning them with plagiarism detection software.

- The researchers need to have one publication in peer reviewed journals before they are awarded PhD degree from an Indian university

CONCERN FOR RESEARCHERS IN INDIA

India is a country with over 1.2 billion people, with more than 65% population below the 35 years of age and a good lot of these youngsters are supposed to involve in teaching and research activities both within and outside the country. With the technological revolution, the IT industry is offering ever now and then has increased the chances of academic misconduct among these budding researchers and academicians. There is every need that these youngsters be made aware about the prevailing academic misconduct across the globe and the need thereof to dissociate themselves from all such immoral and unethical practices. There is a need to understand that any act of plagiarism amounts to

- Academic fraud, which is unacceptable, unethical in the academic world. It is against the rules of academics and publishing.
- It prevents the growth of knowledge and students as critical thinkers.

ACADEMIC MISCONDUCT

Meaning

Conduct is always directly associated with the ethics or the accepted norms in a society about a particular profession or a practice. Moral education is one such aspect of education which is being taught and tried to pass on to the younger generations from their early age. This moral education is mostly about acceptable and unacceptable practices in the society the same is the case with the academic misconduct. Academic misconduct can be termed as those academic practices, which have no legal base to stand on, which are being undertaken by breaching the norms and the ethics of the profession. Some of the common aspects associated with the academic or research misconduct include.

- Fabrication, falsification, or plagiarism in proposing, performing, or reviewing research, or in reporting research results.
- Fabrication is to falsely manipulate data and presenting the results as per ones convenience or suitability.
- The falsification is manipulation of research materials, equipment, processes, or changing or omitting data or results such that the research is not accurately represented or documented in the research record.
- Global problem observed in all disciplines.

Forms

Academic misconduct cannot be limited to one or two forms. There are different forms of academic misconduct and one among them is plagiarism. In the field of academics or research, plagiarism is a practice wherein a researcher or an academician projects the work of others as their own. Perhaps it would be more appropriate to say that plagiarism amounts to intellectual theft. It is more about steeling the work of others and producing, projecting or even reproducing it as one's own.

There is a need to understand that the researchers are always aimed to fill the research gap, and in this process, researchers first try to find out those gaps which exist in the vast pool of information and existing human knowledge about various things. During the course of locating the exiting knowledge or information gaps researchers come across a huge amount of related information in the chosen or given field. And, while carrying forward the research from a particular point there is always a need to justify the existing pool of knowledge by making attributions to all those researchers who may have contributed in this regard. In the process of review there is always a need to properly cite those sources of information maintaining from where a particular piece of information has been taken or even the idea has been borrowed. A researcher cannot simply term the others ideas and research results as his/her own by not acknowledging the original contributor and the moment same is done, the practice amounts to, as an act of plagiarism.

Forms of Plagiarism

These are three common forms of plagiarism. These are:

Plagiarism for Scientific English

This is being done by the scientists or researchers whose native language is not English. The international journals accept manuscripts, which are well written English. Such scientists or researchers are under pressure to get then article paper published in international journals, tend to copy for polished language. This form of plagiarism is seen due to lack of technical writing skill.

Self-Plagiarism

When an author uses his or her exact or precise verbatim words from some previously published work and does not acknowledge the source, it is called self-plagiarism. The practice of self-plagiarism has grown manifold among the researchers all across the world in almost every subject discipline. Researchers have been mostly seen as taking liberty with their writings with the result they presume that they can produce and reproduce their own work at their free will. Researchers and authors need to understand this very basic fact that even if one carries forward ones earlier work, one needs to properly cite his/her earlier work.

Intellectual Theft

Intellectual theft is about copying the text, data or any other form of published material from a particular source without making attributions or acknowledging the actual source or the author. By discrediting the original thought or idea amounts to steeling of research work and academic contributions of actual authors is termed as the act of plagiarism. This practice of unethical academic or research misconduct is also termed as its alright wherein the author usually fails to express his/her understanding about the subject in his/her own words. The result of which is copying and pasting the text from any other relevant source without acknowledging the original author or source of information.

The fundamentals of any research activity are based on crediting those sources of information which may have or which form the basis of any research activity or a research problem. Identifying the research problem from almost already existing knowledge about any subject or object and thereon not acknowledging those sources of information is what forms the basis of intellectual theft or the academic misconduct.

Factors

There are numerous reasons which of late have led to the disproportionate increase in the academic misconduct. Personal and professional growth is the ultimate driving force, which motivates one for greater and quality work and in the process to accomplish more in lesser time and to achieve new heights in lesser time, people generally tend to engage themselves in many undesirable practices to achieve their ends. Some other factors which contribute to the growing academic misconduct among the academia include, the following:

- Enhanced and easy access to Internet
- There are websites, which offer paper writing and other similar services, which amounts to academic misconduct.
- "Publish or perish" syndrome
- References and citations are important in any kind of research writing, or reporting. It has been observed that the inadvertent plagiarism results due to lack of proper referencing skills.
- It has also been observed that the students are often not explained about plagiarism or unauthorized copying (Power, 2009)
- Information Technology has empowered people to access and freely exchange ideas. Information is acquired in few seconds. Earlier information was searched, evaluated and processed in order to synthesize new ideas

- Change in information-seeking behavior of the people in general and academia in particular
- The importance of the library as a place is dwindling (serendipity of information)
- "Squirreling behavior" (Rowlands and Fieldhouse, 2007)
- Other miscellaneous factors: Pressure of deadlines, feelings of incompetence, shortage of time, ambitions and expectations

IMPACT

By indulging in immoral acts in academics and research, a researcher or an academician should never expect of reaping any exceptional dividends from the work, which may be based on falsification, theft, deception and all other ill wills. By indulging into academic misconduct, one may rise to new career highs, but that does not mean the person is worth it. By adopting the practices of plagiarism, one may scale the ladder of success fast with greater ease, but the fact is, one has got nothing to offer to the people to whom one is aimed to serve. The undesirable actions of academic misconduct are bound to harm the public interests to far greater degree than to serve the interests for which one chooses to trade such course. Some other common impacts of academic misconduct involve.

- Most of the research is government funded
- The public money is not being utilized efficiently and judiciously (Saha, 2014)
- Scientists lose credibility among the masses (OECD workshop, 2007) it is well known that fraudulent practices and various forms of misconduct can cause colossal damage to research; tarnish the image of the institutions as well as the country. It also betrays the trust of the public which supports research by paying various taxes. Misconduct especially in applied research in agriculture, veterinary sciences and medical sciences may harm the human and animal lives.
- New knowledge is not generated
- Society as a whole suffers (Lancet article published in 1998)
- Duplicate or redundant publication, which is a kind of self-plagiarism encroaches upon precious journal space, which is much sought after by the researchers.
- The duplicate or redundant publications also waste the limited resources of the editorial and peer review system
- Salami slicing or data fragmentation is an unacceptable practice in science and research because it can distort the literature and the readers may think that the data present in each article has been derived from different sample (Roig, 2006)
- Article retraction: Retraction notice is issued by the editor of a journal to alert the readers of an article, which is no longer valid as it was based on fabricated or false data. The retractions are occurring with amazing frequency and have become so common at international level that a blog has been launched, "Retraction Watch" (http://retractionwatch.com/) to keep a track of all retracted articles and report about them
- Fang, Steen and Casadevall (2012) have found that out of the 2,047 retracted biomedical and life sciences research publications indexed in PubMed, 21.3% were due to error, 68.4% were cases of academic or scientific misconduct, which involved fraud, duplicate publication and plagiarism.

OTHER UNACCEPTABLE PRACTICES

A researcher is always supposed to be bound by the research ethics and those who do not follow their professional or research ethics do the greatest disservice to their profession and their academic community.

Any research activity undertaken by a researcher should always be free from bias and personal prejudice. A researcher should never let the data be manipulated, misrepresented to influence the investigation. **Farthing (2014)** emphasized that there should be no selectivity in data analysis and reporting, inappropriate image manipulation and misreporting of errors are collectively questionable research practices as these affect the research output and research culture adversely.

WHAT CONSTITUTES PLAGIARISM?

It is being observed that both academicians and research scholars appear a bit confused over the subject as what constitutes the plagiarism, as most of the time they defend and justify their argument by supplement that despite acknowledging the original authors, their work is being considered as suffering with plagiarism. Researchers generally quote the text from other sources and paste it in their research paper or chapter as it is without presenting any correction with their own research. All such researchers actually do not want to lose the actual essence of the text taken from other sources, hence paste the borrowed text as it is, but do acknowledge such texts at the end of the chapter or in the reference section. Now the bigger question is, does this practice of copying, pasting and acknowledging the authors tantamount to the plagiarism, and the answer is yes. Unless and until a researcher is not going to reflect his/her personal understanding about the subject, the work can be rated and termed more as a theft. And this is something which both the researchers and the academicians are unable to understand.

Some other common practices which tantamount to plagiarism include:

- Showing someone else's work as your own
- Copying words or ideas from someone else without giving credit
- Failing to put quotations
- Copying so many words or ideas from a source that it makes up the majority of your work whether you give credit or otherwise.

PLAGIARISM DETECTING SERVICES

The similarity check started with open source software, which were mostly freely available on the internet. People interested in checking their text or any other written content for similarity used to download these software and run similarity check. Realizing the potential of the similarity check software, the software developers have come up with proprietary software, which more or less can be termed as the paid services, whereby the institutions need to sign agreement with service providers for availing the service against the payments. The market potential of plagiarism detection software is very huge and wide, as a good number of service providers in the field has already come up with their software and they are doing a brisk business.

Urkund, Turnitin and iThenticate are the three main software available as on date in Indian market. The service providers of the software have already tied up with governing bodies of the Indian higher education system like, UGC to extend these services to various university, colleges and other higher education institutions across the country.

Turnitin (turnitin.com): It is a cloud based service for originality checking, online grading and peer review. It saves teachers' time and provides rich feedback to the students. It compares the text of uploaded draft to a vast database of over 45+ billion pages of digital content.

Over 337 million submissions in the student archive and 130,000+ professional, academic and commercial journals and publications.

iThenticate: iThenticate prevents misconduct by comparing manuscripts against its database of over 34 billion web pages and 129 million content items, including 36 million published works from over 465 scholarly publisher participants of Crosscheck, a service offered by CrossRef and powered by iThenticate software and developed by iParadigms.

Urkund: Urkund is very user friendly and easy to use. The students, scholars and teachers of the institutions who subscribe the software can easily register on the Urkund at its portal, which is accessible at www.urkund.com. The Urkund software is owned and developed by Prio Info Sweden in the year 2000.

Functions of Anti-plagiarism Detection Software

Keeping in view the requirements of the market in general and academic and research institutions in particular, these anti-plagiarism detection software have come up with a variety of features? These software apart from being able to scan the documents for similarity check against millions of documents across the web, they also offer an added advantage to students and teachers to assign and submit their assignments online. Some common functions of these anti-plagiarism software include,

- Originality check
- Engages students
- Manages assignments
- Improves faculty evaluation process
- Provides rich feedback
- Promotes academic integrity check.

Advantages of Anti-plagiarism Software

Some common advantages of anti-plagiarism detection software are

- Quality of assignments is increased
- Instant feedback from the teachers
- Students learn from one another through peer review
- Citation and referencing skills are enhanced
- Acts as a deterrent for the students and researchers.

Limitations of Anti-plagiarism Software

Advantages and disadvantages are like the two faces of the same coin. There may hardly be any system or scheme of things in place, which may only show the presence of advantages and absence of disadvantages. Accordingly, if these plagiarism detection software on the one hand offer us a series of advantages, on the other hand these software do have their own limitations. Some common limitations, which these software suffer with, include

- Plagiarism detection software can only scan those documents which are available in digital form. This means any content or material taken from any printed document will be excluded during the scan process for obvious reasons.
- Similarity checks are performed by these software only against those documents which are available online. This also means that digital documents not available online will be excluded during the scan.
- The service providers of these software are required to sign MoU with various online publishing houses who maintain a database of their publications, which are available and accessible to the users online, no matter if, on payment basis. These service providers are required to pay for accessing the databases hosted on different web servers by various publishers across the world. This also means that it is not

necessary that all the online documents are scanned for similarity check unless access is not granted for this check by the respective publisher. Besides, the fact also remains that online publishing houses may charge at exorbitant rates for providing access to their databases and any inability to pay the required fee can easily exclude all such databases from scan with whom understanding may not reach.

- Format of text and language is also the other added problems associated with the online scan, as these software pick up that text format and those languages with which they have been customized. Any alien language and file format means it stands excluded from the scan.

- There are also some customization problems with these software as they sometimes do not reflect the copied content.

Based on above mentioned facts, while scanning the documents with the help of these software there is a far greater need that same document be checked or cross checked manually by the subject specialist to assess and analyze the amount of seriousness reflected by the plagiarism detection software. It is not always necessary that the content, which the software may detect under similarity parameter, may necessarily fall under the purview of plagiarism.

Word of Caution in Use of Anti-plagiarism Tools

Some arguments and counterarguments are being made both in support and against the true functioning of these plagiarism detection software. Firstly, there is need to understand the basic fact, that it is the artificial intelligence which IT industry is making use of and there on drawing comparison between the human and artificial intelligence may sound improper. Artificial intelligence always works on a predefined path, while as the human intelligence is more about exploring newer possibilities with mental cognition.

- The software cannot be substituted for human intelligence, the supervisor is the best person to judge the true context of plagiarism, if any
- The instructor has to take the final call
- Support or aids in order to give effective feedback to the students
- Tools teach the students about the best practices of scholarly writing and communication.

ROLE OF LIBRARIES IN COMBATING ACADEMIC MISCONDUCT

Since the inception of the formal education system, libraries have been passively involved with the academic and research activities of their respective institutions. It has been observed that libraries as institutions and library professionals as passive teachers play a very significant role in accomplishing research projects? The library professionals have been found as playing a very significant role in curbing this academic misconduct. Some of these key role include,

- The role of libraries comes into play from the very first step of the research activity and that is formulating the research problem. Workshops can be organized in the research process, covering the aspects like, how to formulate research questions and how to locate and evaluate sources
- Research writing is both art and skill and this skill can be easily honed among researchers by organizing special training workshops on writing research papers, review writing, article writing, paraphrasing, referencing, citations, etc.
- Workshops on plagiarism can be conducted at regular intervals of time among the teaching and scholarly community of the institution, wherein attempt should always be made to draw awareness among the researchers and sensitize the scholarly community about the repercussions if found guilty of violating research ethics or been found involved in the acts of plagiarism.

- Awareness can be drawn about the ways and means of consulting different sources of information, be the print, electronic, online, offline or any other sources which are available and accessible to their users.
- Awareness about acknowledging sources, use quotations, paraphrases, and citation styles and avoid copying can be drawn.
- Following teaching and research ethics

ROLE OF TEACHERS/SUPERVISORS /GUIDES

The teachers should act as mentors to clarify and demonstrate with examples the rules of academic conduct. When the researchers and faculty members make imprudent decisions because of stress, time pressures and inadequate mentorship, it all results into misconduct. The stress may relate to the desire to get timely promotions, grants, inadequate understanding of research processes. Saha (2014) has appropriately mentioned that, "a little hand holding of researchers would pay rich dividends in nurturing science and technology in India."

BIBLIOGRAPHY

1. Carroll J, Oxford Centre for Staff Development. A Handbook for Deterring Plagiarism in Higher Education, Vol. 2. Oxford: Oxford Centre for Staff and Learning Development; 2007.

2. Eret E, Gokmenoglu T. Plagiarism in higher education: A case study with prospective academicians. Procedia-Social and Behavioral Sciences. 2010;2(2):3303-7.

3. Fang FC, Steen RG, Casadevall A. Misconduct accounts for the majority of retracted scientific publications. Proceedings of the National Academy of Sciences. 2012;109(42):17028-33.

4. Farthing MJ. Research misconduct: a grand global challenge for the 21st century. J Gastroenterol Hepatol. 2014;29(3):422-7.

5. iThenticate (2012). 2012 Suvery highlights: Scholarly Plagiarism. [online] Available from https://www.ithenticate.com/hs-fs/hub/92785/file-16017086-pdf/docs/plagiarism-survey-results-120412.pdf   on [Accessed July 01, 2016].

6. Larkham PJ, Manns S. Plagiarism and its treatment in higher education. Journal of Further and Higher Education. 2002;26(4):339-49.

7. Power LG. University students' perceptions of plagiarism. The Journal of Higher Education. 2009;80(6):643-62.

8. Retraction watch (n.d). Retraction watch. [online] Available from http://retractionwatch.com/

9. Roig M. Avoiding plagiarism, self-plagiarism, and other questionable writing practices: A guide to ethical writing. 2006.

10. Rowlands I, Fieldhouse M. Information behaviour of the researcher of the future: Work Package I: Trends in Scholarly Information Behaviour. 2007.

11. Shafer SL. You will be caught. Anesthesia & Analgesia. 2011;112(3):491-3.

12. Saha R. Plagiarism, research publications and law. Current Science (00113891) 2017;112(12):2375-8.

13. Sowden C. (2005). Plagiarism and the culture of multilingual students in higher education abroad. ELTJ2005;59(3):226-33.

14. Turnitin User Guide (n.d). Available from https://guides.turnitin.com/01_Manuals_and_Guides/Student_Guides/01_QuickStart_Guide

15. University Grants Commission (2009). UGC minimum standards and procedure for award of MPhill and PhD degrees regulation. [online] Available from http://www.ugc.ac.in/oldpdf/regulations/mphilphdclarification.pdf on [Accessed December 01, 2016].

Chapter 3

Plagiarism Detection Tools

INTRODUCTION

There seems to be no end to the technological offings. Each new day the technology comes up with new application software and new tools, which have enabled men of the present day world to push beyond limits. Plagiarism detection software have become the buzz word these days among the researchers and academia across the world. With the use of these technological offerings, the entire academia and the research community are able to assess the quality and originality of their research work. Plagiarism detection software's have given a new lease of life to the quality research, which otherwise of late suffered to a considerable level due to academic theft or academic misconduct, what is generally termed as an act of plagiarism. In the present chapter, an attempt has been made to discuss about the three plagiarism detection software, Urkund, iThenticate and Turnitin, the services of which are being availed by universities and other higher education institutions across the world. In India the University Grants Commission has extended services of Urkund software free of cost for plagiarism detection service to universities recognized under 12 (B) and 2 (f) act of the UGC. Accordingly, in the present chapter we shall be talking about various features of these software, how to make use of them and the facility to create assignments by the teachers on these software, especially by pressing the services of TurnitIn

Examples

* Turnitin.com http://turnitin.com/
* iThenticatehttp://www.ithenticate.com/
* Urkund http://www.urkund.com

TURNITIN

* Web-based service to manage the process of submitting and tracking papers electronically, providing better and faster feedback to students.
* Text matching
* Turnitin uses three databases for content matching:
* 45+ billion web pages crawled
* 337+ million archived student papers
* 130+ million articles from 110,000+ journals, periodicals & books
  www.turnitin.com

English, Arabic, Chinese (Traditional and Simplified), Dutch, Finnish, French, German, Italian, Japanese, Korean, Polish, Portuguese, Romanian, Russian, Spanish, Swedish, Turkish and Vietnamese.

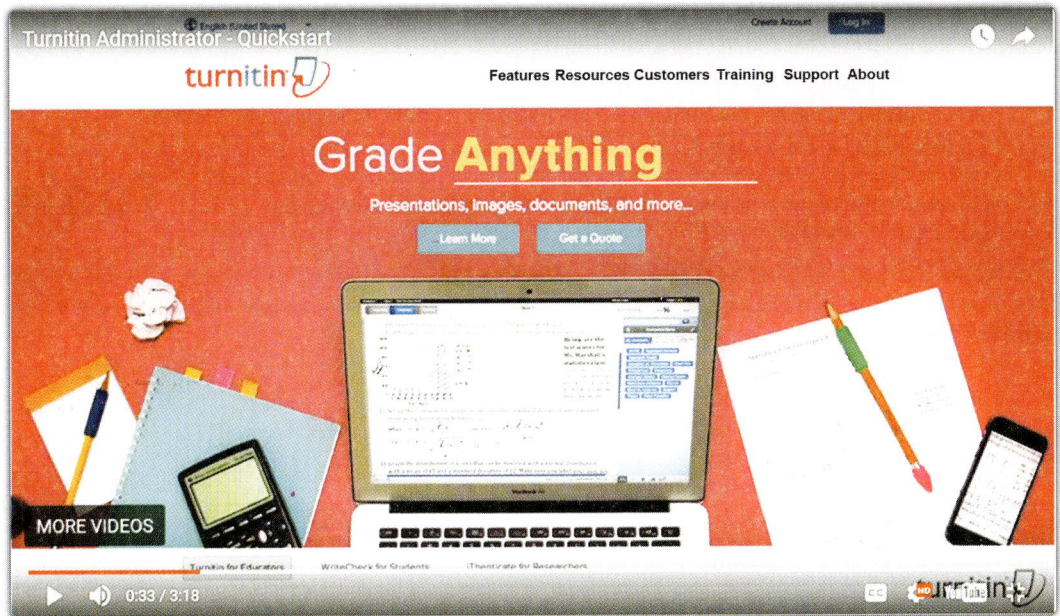

Fig 1. The main window of turnitin

WHAT DOES IT DO?

- Originality Check
- Engages students
- Manages Assignments
- Improves Faculty evaluation process
- Provides rich Feedback
- Promotes academic integrity check

HOW DOES IT WORK?

Universities and other institutions can subscribe to plagiarism detection software on a yearly basis by entering into an agreement with the service provider by working out different modalities given the suitability of both service providers and the one who avails the services.

- A university administrator creates accounts for faculty members.
- Faculty members set up classes and assignments for their students.
- Faculty members enroll students in their classes or they may enroll by themselves.
- The students submit their assignments through Turnitin, check their originality.
- The faculty members may evaluate the submissions give feedback, write comments in the drafts.
- It tells /highlights the content which matches with the already published online content.
- It does not decide if plagiarism has occurred. The faculty has to see closely/minutely and decide if plagiarism has occurred.

ADVANTAGES

- Prevents Plagiarism
- Engages students (engages the students to make concerted efforts to improve the report).
- Students can get instant feedback.
- Peer review (teachers can let their students anonymously critique and evaluate each other's papers).
- Identifies the different words which have been added, deleted, or substituted.
- Does citation verification.

Instructors as well as students can upload papers

DISADVANTAGES

- Cannot identify plagiarism from a non online source.
- Has problems with mathematical formulas (latex files).
- Distorts the format of the original documents: tables, graphs, and images don't appear.
- Does not differentiate between quoted materials and original writing at times.

WAYS TO USE TURNITIN.COM

- Extensive use
- Sparse use
- Log in to
- Change temporary password
- Creation of classes

Creation of assignments

Using Turnitin

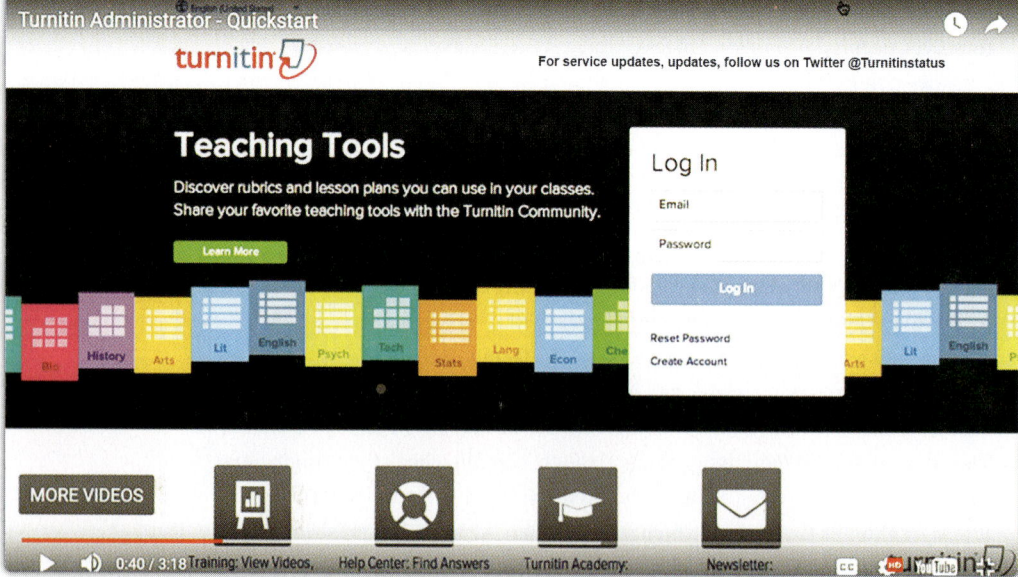

Fig 2. The main login window of the turnitin

FOR TEACHERS

- Log into
- If you have forgotten your username
- Setting up your class
- Click on the blue tab, "Add a class"
- Select an enrolment ID, PW
- Select class, start date and end date
- Class ID numbers and Passwords

CREATING ASSIGNMENTS

- After you set all parameters you submit.
- You will now see that assignments under the appropriate class.
- You may continue adding assignments

FOR STUDENTS

- Teachers will create accounts for their students.
- They will receive e- mail notification.
- Use the password to log in and may change their passwords.
- Click to submit/ upload any file
- Students submit assignments.
- Important to note that each submission corresponds to each assignment.
- If you intend to use TII sparingly:
- Log in to
- Change password given to you.
- Enable Quick Submit
- Submit papers

EXTENSIVE USE

- Getting started for teachers
- Log into with UN and PW which has been communicated to you by the administrator.
- You can change your password
- If you forget your password, it can be recreated by clicking on "forgot password".

Create class by clicking on, "add class" Tab.

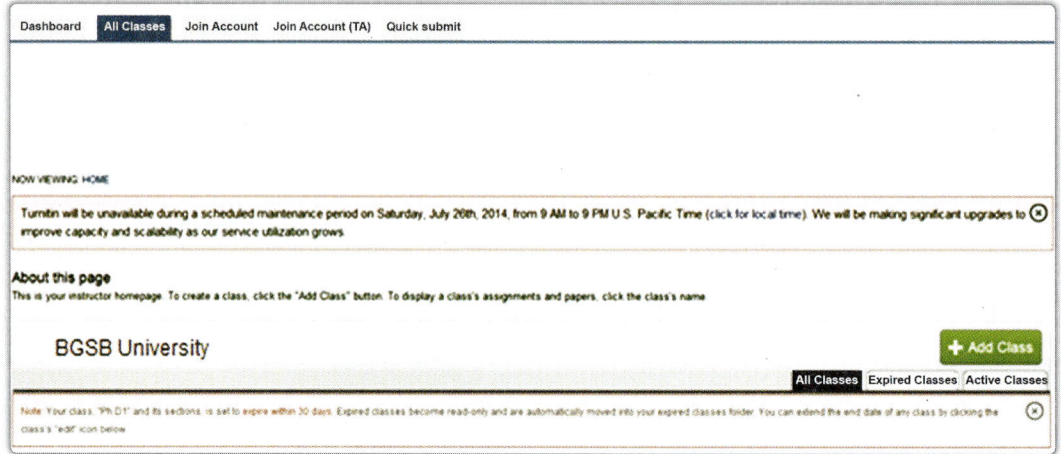

Fig 3. Winodw to add new classes

Create Class

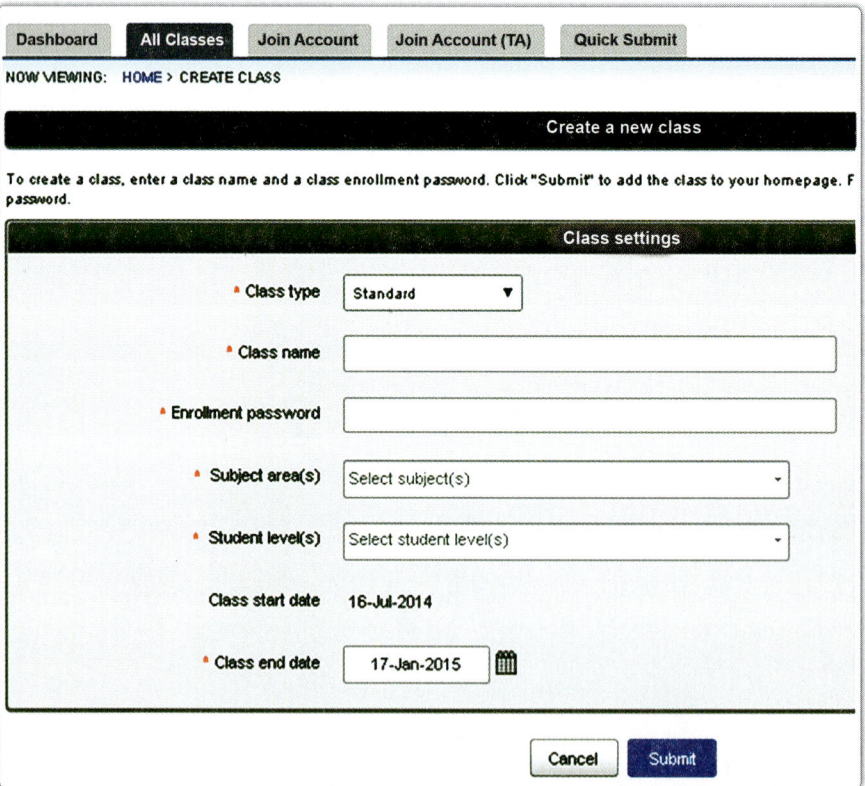

Fig 4. Window to add new classes-II

"All classes" Tab

- If you click on, "All classes" Tab, the following screen will be displayed

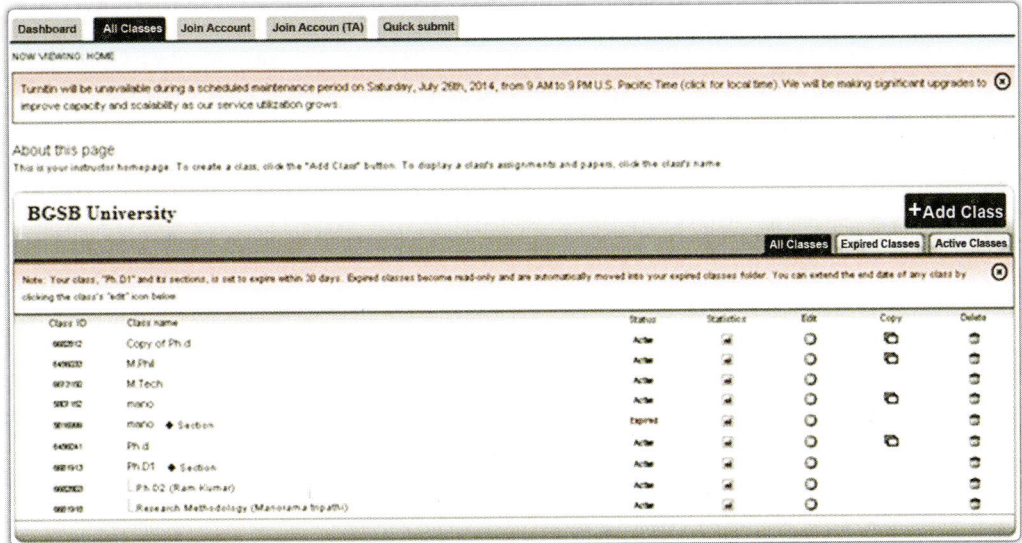

Fig 5. Window to view all the created classes

ADD ASSIGNMENT: Instructors can add assignments, set different parameters and then click submit. Once the assignments have been created, the students can upload their drafts. The instructor can submit the assignment for the student who is not enrolled.

- Instructors can add assignments
- Select Assignment type
- Paper
- Peer
- Revision
- Reflection
- Set different parameters and then click submit

Submission of Assignments by Students: You can submit papers to assignments which have been created by the instructor. To submit a paper using file upload, please do the following

- Log into your account.
- Click on the title of the Class you wish to submit to.
- Click on the blue "Submit" button for the assignment you wish to submit to. If there is a grey "Submit" button, submissions do not allow for this assignment. Please check the assignment start and due dates and the assignment info icon.
- Choose Single File Upload.
- Enter title of your paper.
- Click on "browse" to locate the paper saved to your computer.
- Click on the file and click "open".
- Click the "upload" button at the bottom.

- **STOP, and WAIT** for the next page to appear.
- Click "submit" to confirm your submission.
- After you successfully submittal digital receipt will be displayed on the screen. The digital receipt has a paper ID, which is confirmation that Turnitin has received your submission.
- Students and instructors receive a digital receipt for every paper/assignment/draft they upload.

FORMATS OF ASSIGNMENT SUBMISSION

- Microsoft Word® (.doc / .docx)
- OpenOffice (.odt)**
- WordPerfect® (.wpd)
  PostScript (.ps/.eps)
- Adobe® PDF
- HTML
- Rich text format (.rtf)
  Plain text (.txt)
- Minimum of 25 words must be under 20MB not more than 400 pages.

ASSIGNMENTS TO BE ADDED BY THE INSTRUCTORS, WINDOW-I

- Under a particular class, there is a green tab of "Add Assignment"

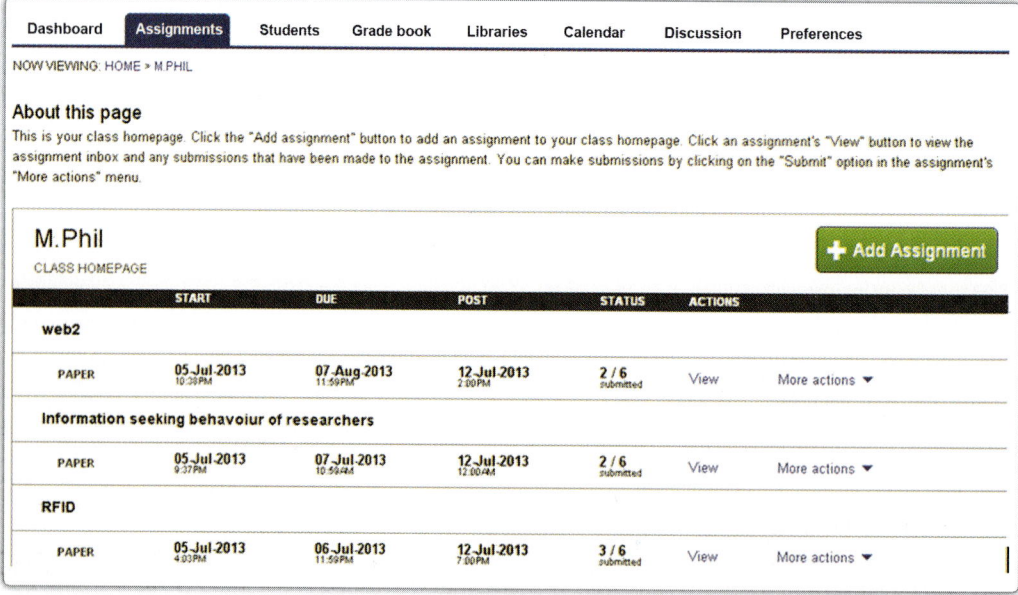

Fig 6. Window to choose ones class to submit the assignment

Assignments to be added by the instructors, Window-II

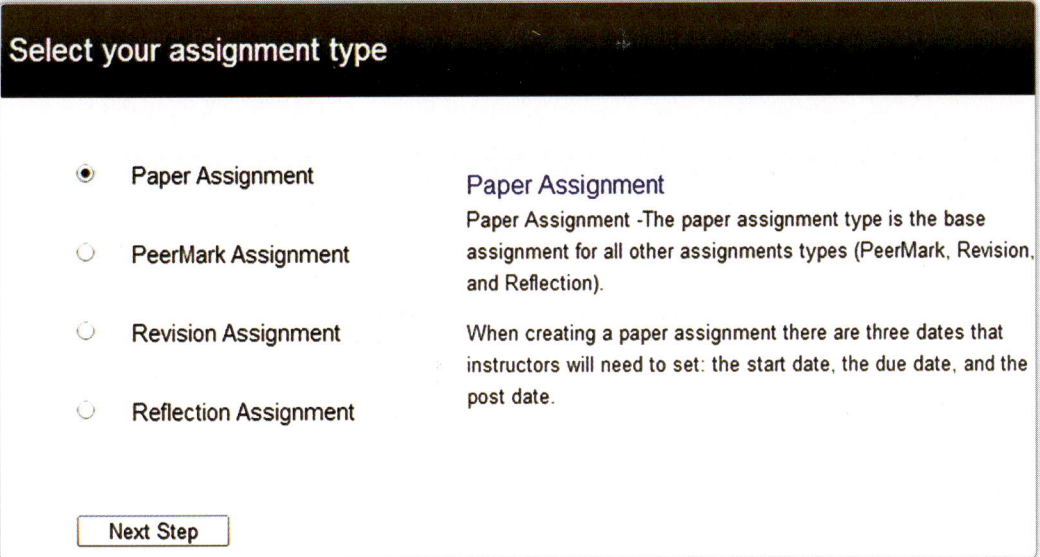

Assignments to be added by the instructors, Window-III

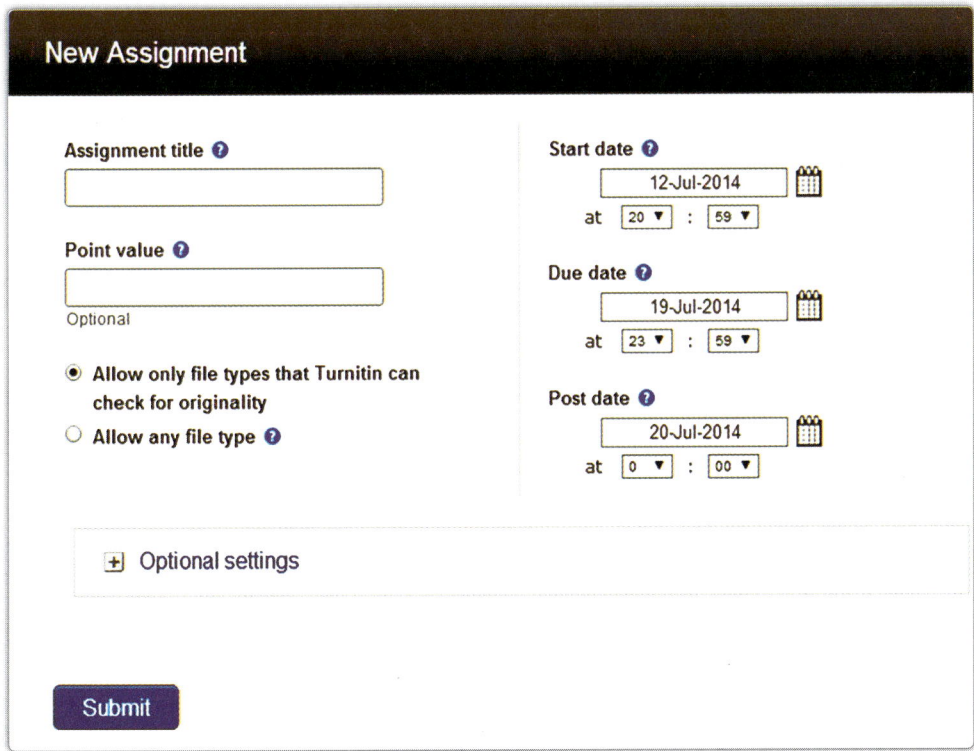

Window to check or changes settings before submitting document for similarity check

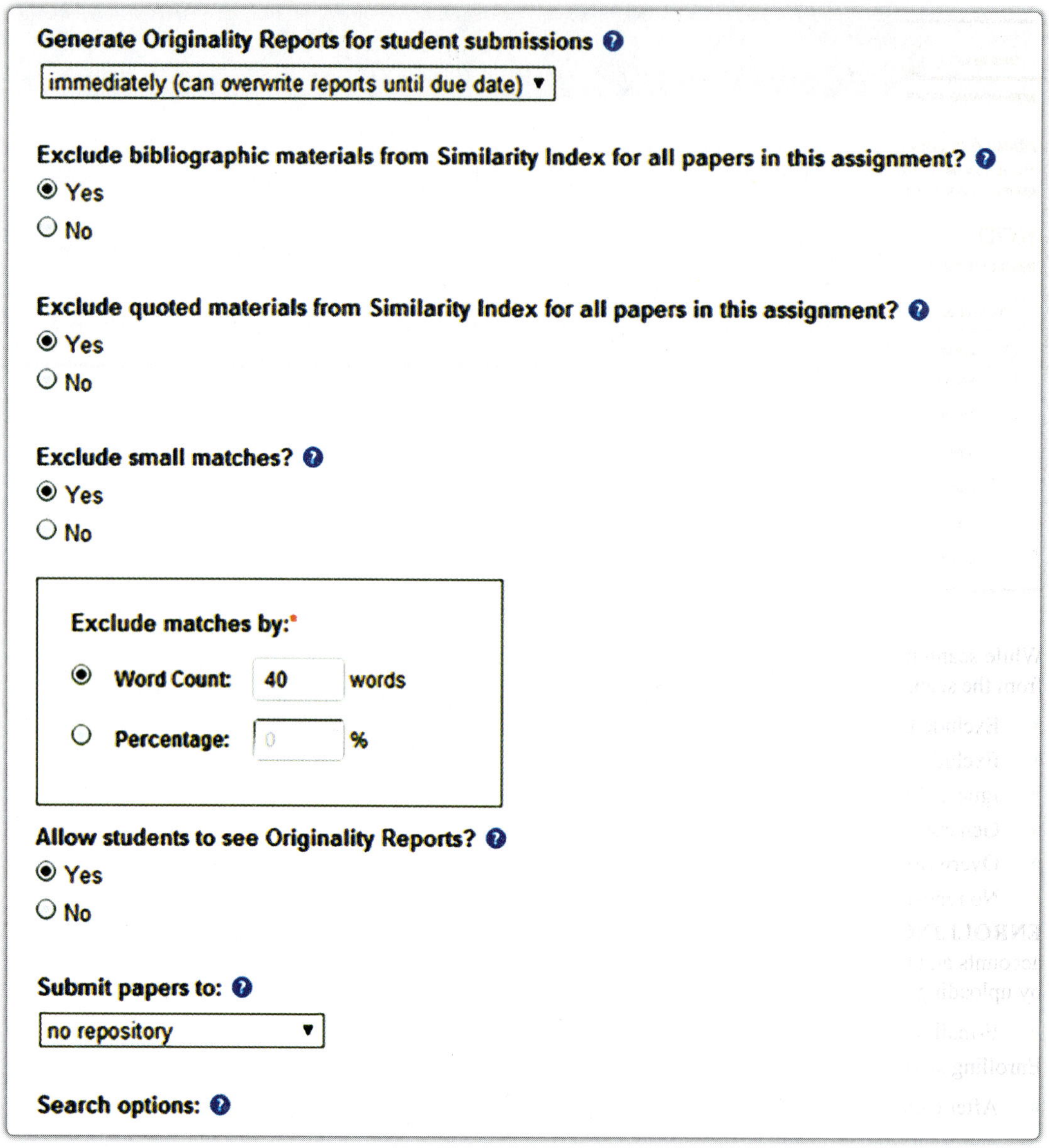

Generate Originality Reports for student submissions ❓

immediately (can overwrite reports until due date) ▼

Exclude bibliographic materials from Similarity Index for all papers in this assignment? ❓
◉ Yes
○ No

Exclude quoted materials from Similarity Index for all papers in this assignment? ❓
◉ Yes
○ No

Exclude small matches? ❓
◉ Yes
○ No

Exclude matches by:*

◉ **Word Count:** 40 words

○ **Percentage:** 0 %

Allow students to see Originality Reports? ❓
◉ Yes
○ No

Submit papers to: ❓

no repository ▼

Search options: ❓

Assignment Inbox

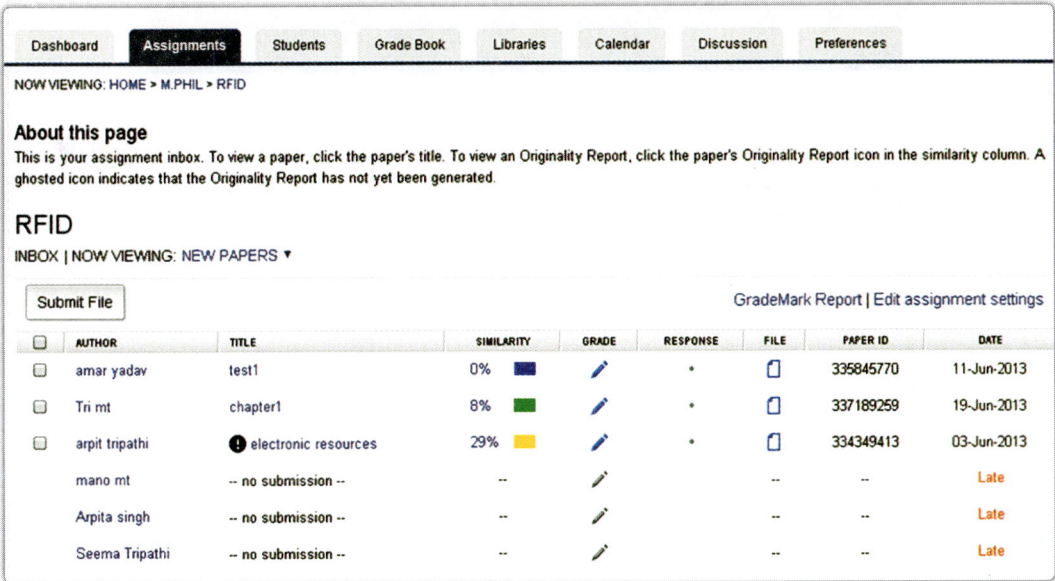

While scanning the document for plagiarism, a teacher can set parameters so as to be included or excluded from the scan. some of the parameters which can be set as per ones preference include.

- Exclude Bibliography
- Exclude quotations
- Ignore-40/50 words
- Generate OR
- Overwrite reports
- No repository

ENROLLING STUDENTS: Instructors submit students' papers/assignments, instructors create students' accounts and they submit their assignments. The students can be enrolled by clicking on "Add students" tab by uploading students list.

- E-mail notification to all the students can be sent.

Enrolling students for the class

- After creating a class, an instructor has to enroll students.

Add Students

Enroll a Student

To enroll a student, enter a first name, last name, and an email address and click submit.

If the student already has a Turnitin user profile, they will be notified and enrolled in your class immediately. If they do not have a profile, we will create one and send them an email notification with a temporary password.

Add student to

Class name: BGSB University, M.Phil

First name

Last name

Email (User name)

Submit

Upload Student List

Upload Student List

Choose a file to upload:

Choose File   No file chosen

File Formatting Guidelines

Your file can be in either **Word, Excel, or plain text** format. For each user in your list, you must include the user's first name, last name, and e-mail address in this order:

   first name, last name, email address

More info...

Upload List

Students create their own account

- Click on "create Account"
- Let the student create a user profile
- Share your Class ID and password with the student.
- Thus, they create their own account and submit their work.

Create a New Student Account

Class ID Information

All students must be enrolled in an active class. To enroll in a class, please enter the class ID number and class enrollment password that you were given by your instructor.

Please note that the password and pincode are case-sensitive. If you do not have this information, or the information you are entering appears to be incorrect, please contact your instructor.

Class ID

Class enrollment password

User Information

Your first name

Your last name

"Statistics" Tab

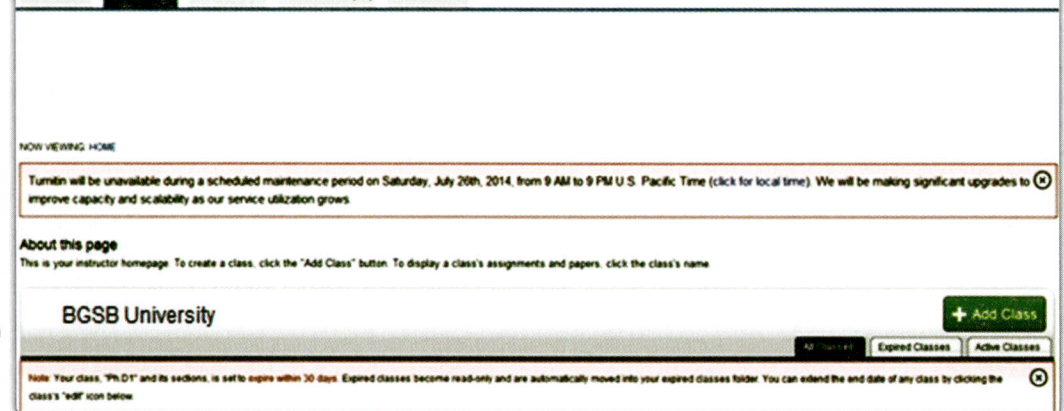

Name	JoinEnrollment password	ID	Students	Submissions	Originality Reports	75-100%	50-74%	25-49%	0-24%	No matches	Peer reviews	GradeMark	Graded papers	Discussion replies	Discussion topics	QuickMark breakdown
M.Phil	chennai1	6456033	6	122	112	8	7	28	56	13	-	5	-	-	-	view
web2		20620146	-	2	2	1	-	-	1	-	-	4	-	-	-	view
Information seeking behaviour of re...		20868316	-	2	2	-	-	-	-	2	-	-	-	-	-	view
RFID		20875624	-	3	3	-	-	1	1	1	-	-	-	-	-	view
Assitive technologies		21147270	-	1	1	-	-	1	-	-	-	-	-	-	-	view
ETDs		21189540	-	1	1	-	-	-	1	-	-	-	-	-	-	view
Clouds		21189602	-	1	1	-	-	-	1	-	-	-	-	-	-	view
varaious 1		21189611	-	1	1	-	-	-	1	-	-	-	-	-	-	view
Copyright issues1		21189614	-	1	1	-	-	-	-	1	-	-	-	-	-	view
languages		21207813	-	1	1	-	-	-	-	1	-	-	-	-	-	view
German studies		21207830	-	1	1	-	-	-	1	-	-	-	-	-	-	view
Literature		21207835	-	1	1	-	-	-	-	1	-	-	-	-	-	view
Literature		21207836	-	1	1	-	-	1	-	-	-	-	-	-	-	view
GSB		21207983	-	1	1	-	-	-	-	1	-	1	-	-	-	view

Using Sparingly

- Activate, "Quick submit"
- If the instructors want to check papers occasionally, they should enable this feature. Unless or until this feature is enabled, the tab will not show in the instructor' account.
- Activate by clicking on "UserInf" tab

Quick Submit

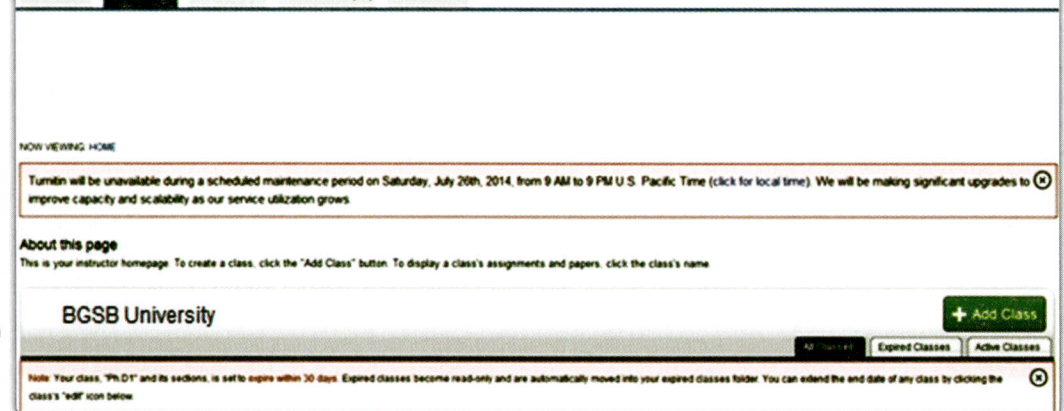

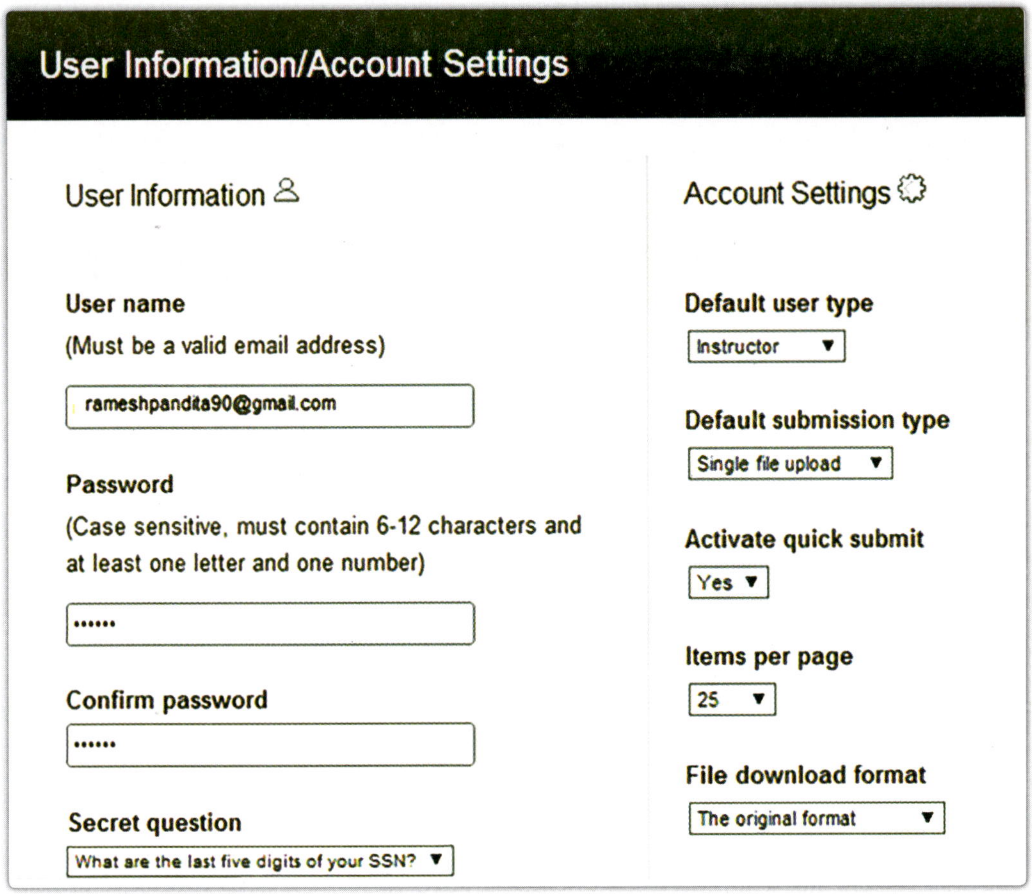

User Information/Account Settings

User Information &

User name
(Must be a valid email address)

| rameshpandita90@gmail.com |

Password
(Case sensitive, must contain 6-12 characters and at least one letter and one number)

| •••••• |

Confirm password

| •••••• |

Secret question

| What are the last five digits of your SSN? ▼ |

Account Settings ⚙

Default user type

| Instructor   ▼ |

Default submission type

| Single file upload   ▼ |

Activate quick submit

| Yes ▼ |

Items per page

| 25   ▼ |

File download format

| The original format   ▼ |

Quick Submit

OR

- It returns colour coded OR to the students or the instructors. A % shows the extent of possible content matches.

It displays the % of similarities based on the following:

- **Blue**
- **Green** 1 -24% of Similarity
- **Yellow** 25 – 49% of Similarity
- **Orange** 50 - 74% of Similarity
- **Red** 75 – 100% of Similarity

ORIGINALITY REPORT

- If the Icon is a grey box with dashes, please allow additional time for the report to be generated.

- If there is no colour coded boxes, but there is text that reads "not available" then the instructor has *not allowed* students to view the Originality Report. Students need to speak with their instructor regarding any of these topics.

UNDERSTANDING ORIGINALITY REPORT

- Originality Reports provide a summary of the matching text found in a submitted paper. The percentage indicates the overall similarity index of the paper, based on how much matching text was found. The highlighted text in the document is colour coded and numbered to match the sources on the right.

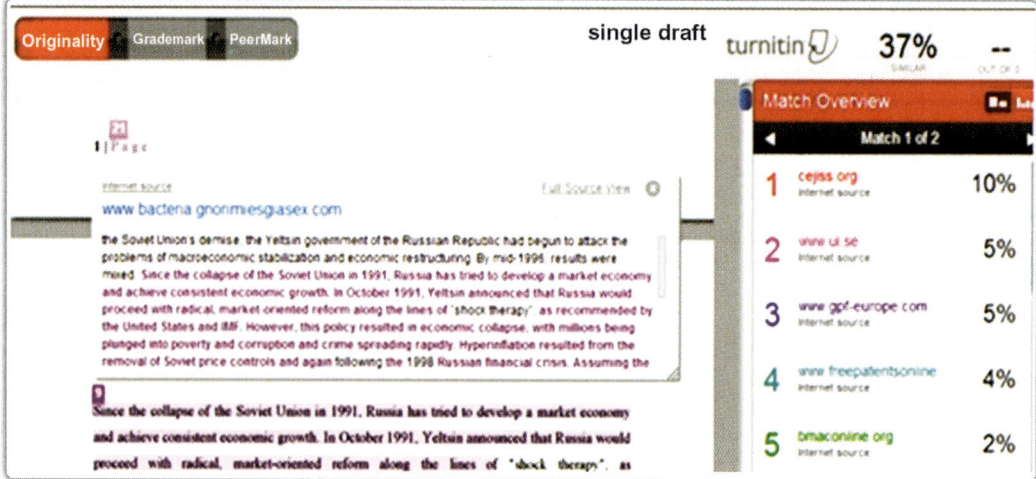

- Turnitin does not have a guide to what is a good or bad percentage. It all depends on the assignment type and what threshold the instructor has set. Research papers tend to have a higher percentage since there is more content from other sources.
- OR can be downloaded and printed as PDF

Excluding Bibliographies, Quotations Etc. from O.R. (Originality Report)

- Open the Originality Report for a submission (under the "Similarity" header, click on the percentage or the colour coded boxes). The Originality Report will open in a new window.
- Click on the "Filters and settings" icon (the icon looks like a small funnel). This is the second icon located at the bottom right of the Originality Report.
- Click on apply changes.
- After applying the new filters, the similarity index (percentage) may change. Any material that is excluded, in this way from a student account **will NOT be saved**. The similarity index of the report will return to its original percentage after the student closes the report. Only changes made by the Instructor get saved.

WARNING: OR

- If a student is quoting from well known sources and same quotation appears in other students' papers or elsewhere on the Internet, you may see an OR for that draft suggests that it has been copied.
- You can "exclude" source by clicking on the "X" icon given on the right of an outside source URL address.

Excluding Sources

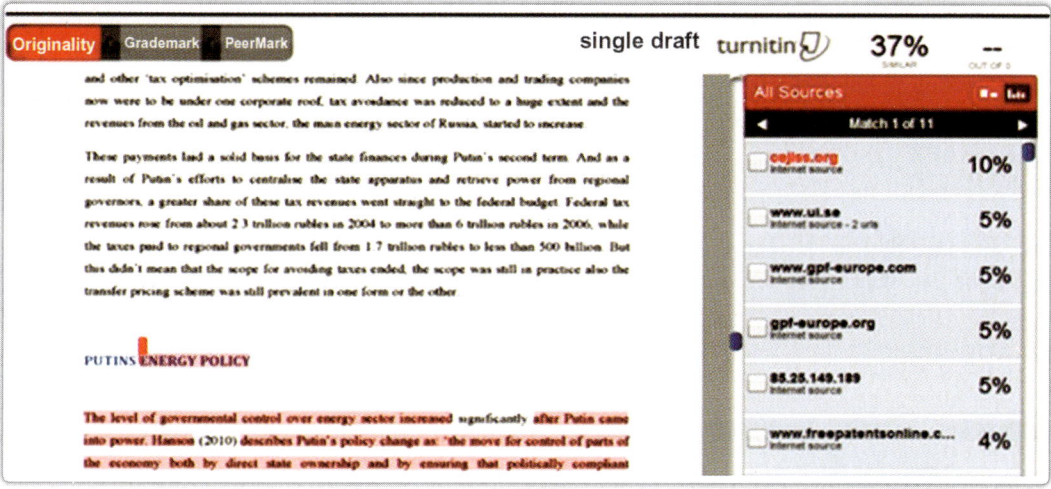

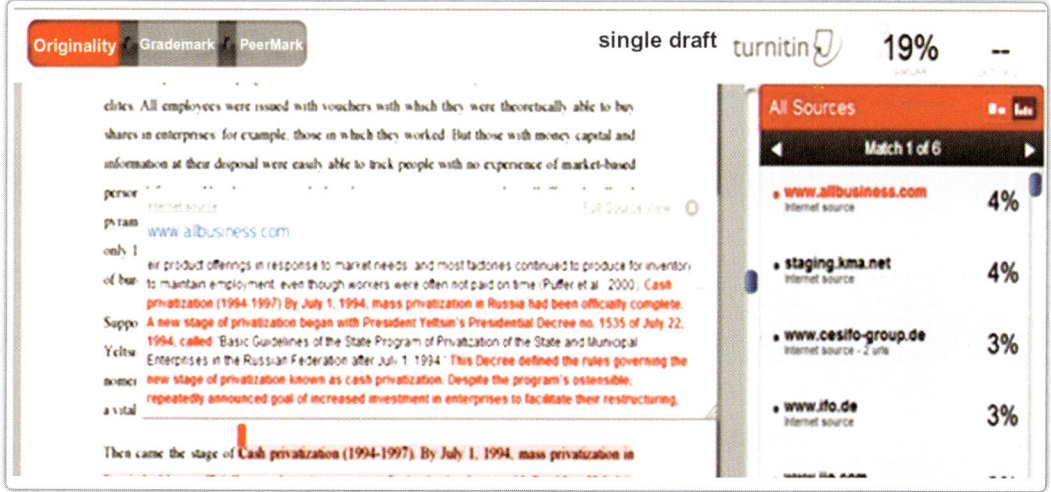

EXCLUDE SOURCES

- Open originality report
- Click on, "All sources" tab given on the top (extreme right) in the right pane
- Select sources by checking the boxes on the left of sources listed.
- Click on "Exclude" tab which turns red after you check the sources to be excluded.

OR (Content taken from Multiple Sources)

- The content which has been copied is colour coded. If the student has used material from multiple sources, each source will be of a different colour. One can click on the colour coded material in the student draft and see the borrowed material in the same colour to the right of the paper/draft.

PEER MARK

- Peer Mark assignments allow students to read, review, and a score or evaluate one or many papers submitted by their classmates. At the end of the Peer Mark assignment, the papers will be distributed so that all the students are able to read the comments left on their work. It may be anonymous or attributed.

GRADE MARK

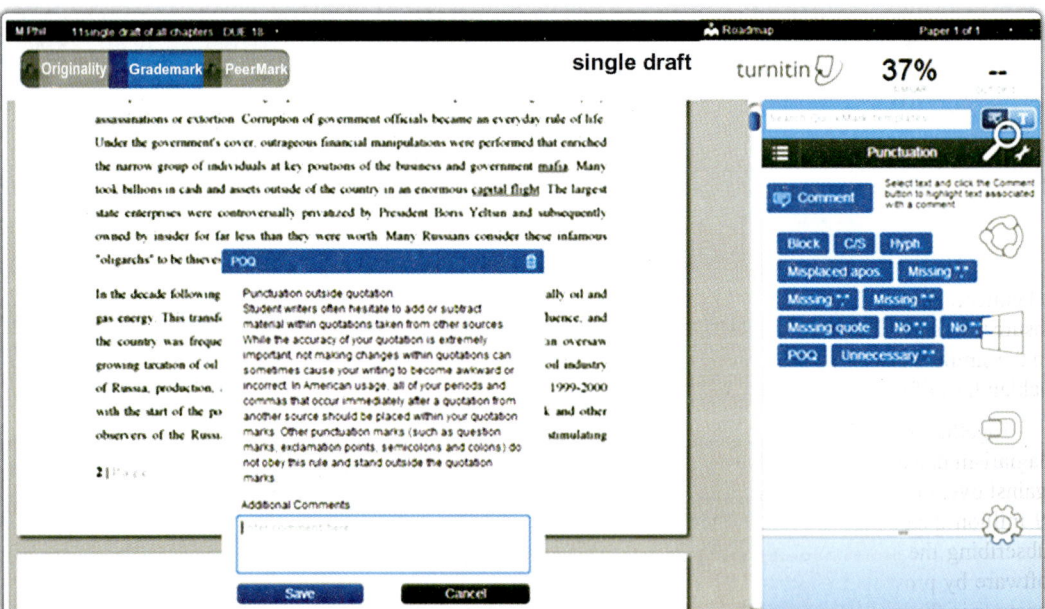

- These tools are not instructors or library professionals, but **support or aids** for instructors or library professionals who need to examine the results of OR carefully in order to give effective feedback to the students

- Do not specify the level of plagiarism, but identify points where the text matches with the published resources.

- Use of ORs under proper guidance-its careful exploitation and interpretation immensely improves the final term paper or research output.

Getting Help

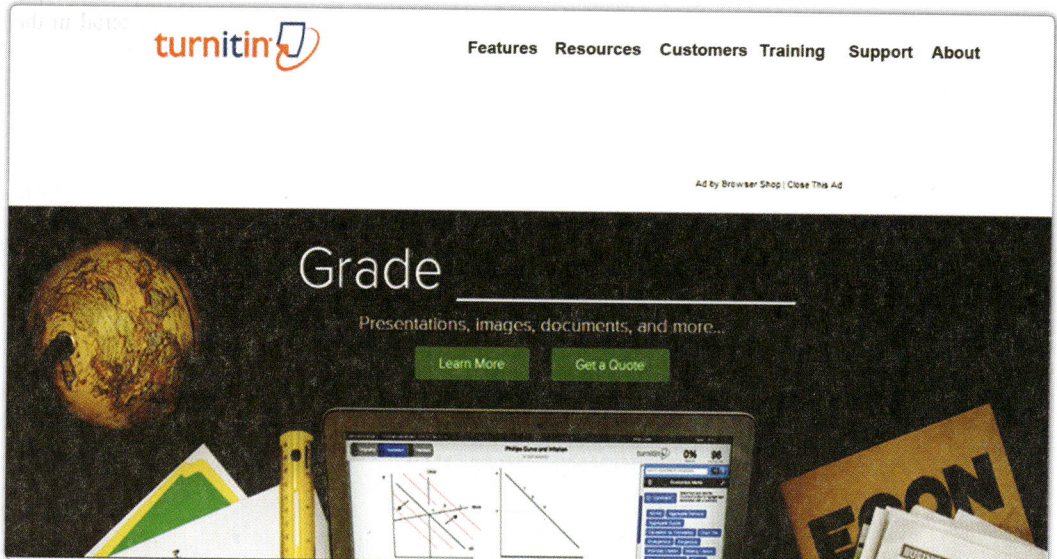

iThenticate

iThenticate is another plagiarism detection tool or software, which is designed to be used by researchers to ensure the originality of the written Work / manuscript before submitting it for publication. iThenticate works on the similar lines to that of Turnitin, in fact iThenticate is developed by the Turnitin with its headquarters in Oakland, California. The company owns its international office in Newcastle, United Kingdom.

Together Turnitin and iThenticate are the leading service providers across the world in the field of plagiarism detection. The company maintains that each document submitted for similarity check is scanned against over 60 billion web pages and over 155 million contents. The company maintains a database of over 49 million documents from over 800 scholarly publisher participants of Crossref Similarity Check. After subscribing the services from the authorized agencies, the users need to first register on the portal of the software by providing necessary information, especially by having one admin, who is authorized to create accounts in the name of other institutional members.

Administrator of the software can create different departments so as to segregate the uploaded documents of each individual department submitted by each individual user from the respective department. Each individual user has to upload documents as per the permission granted by the admin, which may get archived at a specific location.

ACCESS TO ITHENTICATE

- All UGC-funded universities were initially provided with iThenticate access, but later on UGC tied up with another agency (Urkund) for similar kind of services hence, the access to iThenticate was discontinued thereafter.
- The first step is to request the creation of a personal iThenticate account.
- The Librarian is generally the administrator for this software and may be requested for creating an account.
- Students can check documents-assignments, articles and book chapters that they write and which they need to for evaluation, publication, grant proposals, theses etc.

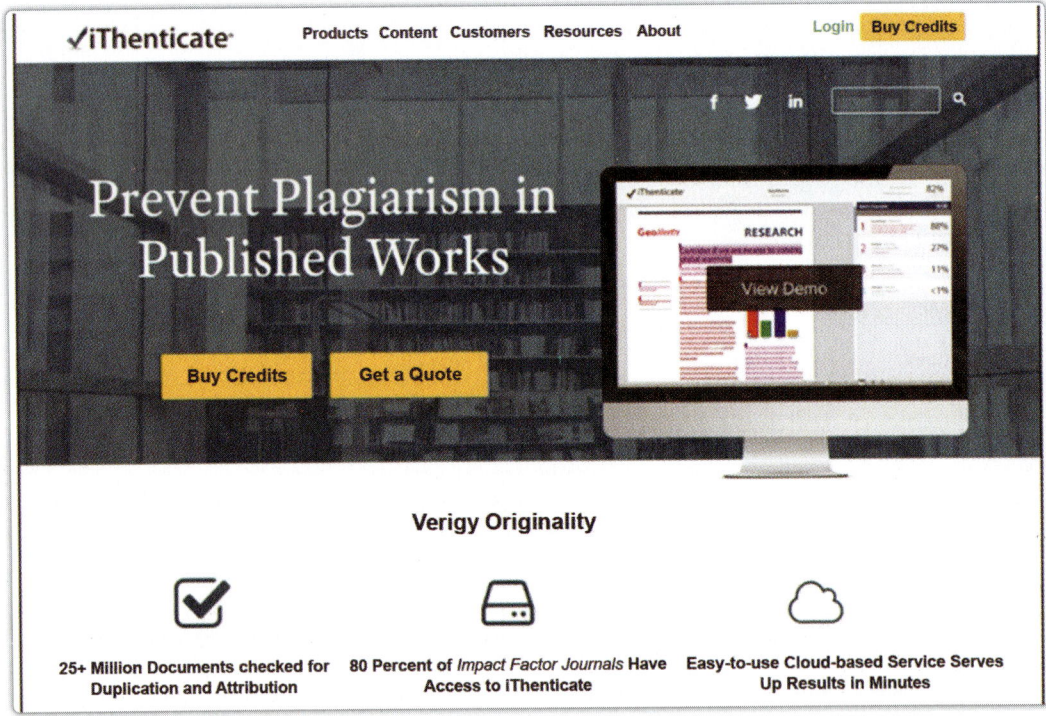

Fig 7. Main screen of the iThenticate window

To submit a document for similarity check, registered users need to login to the portal and then upload a file from the system

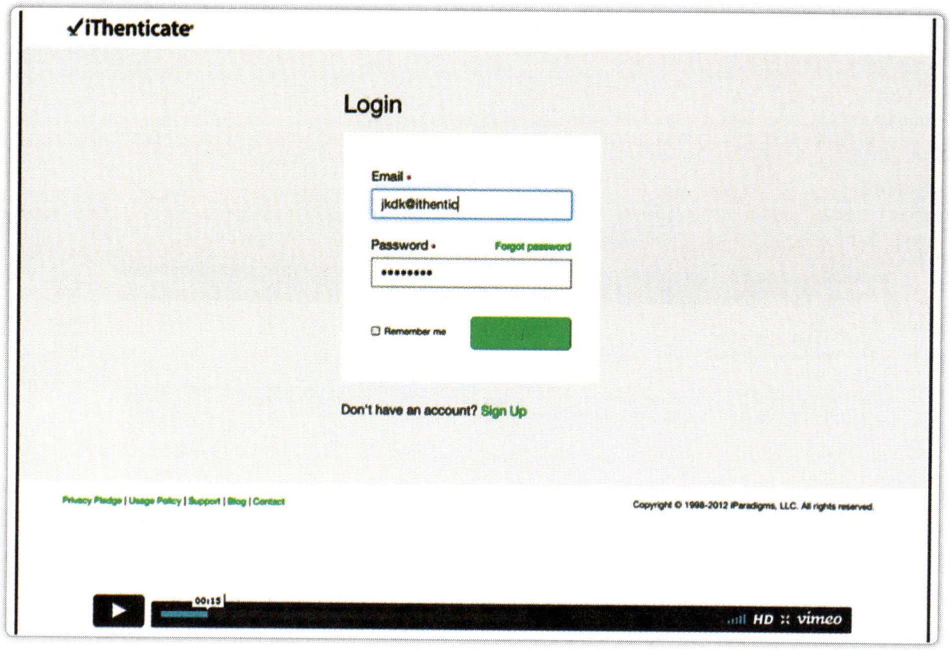

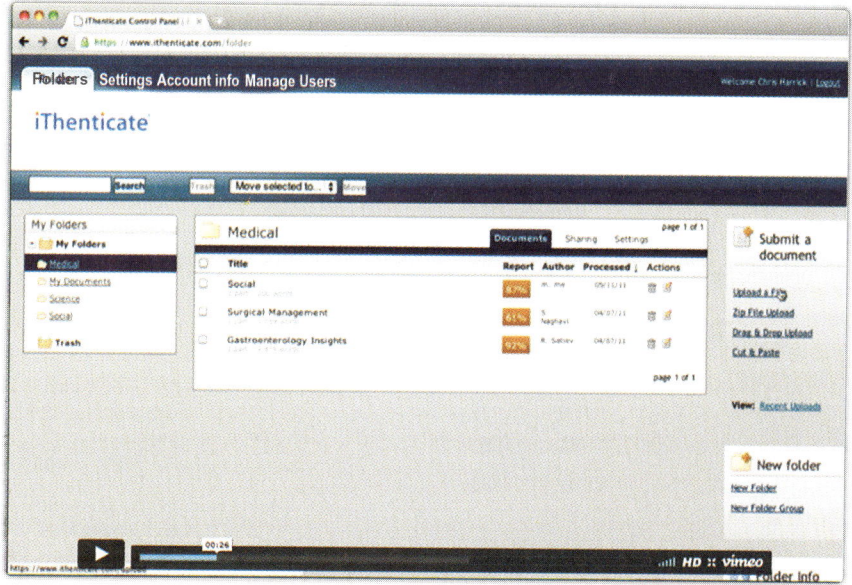

Fig 8. Login window for users

Necessary information about the document is to be provided so as to particularize each individual document. The necessary information to be provided about the documents include, title of the document, author information, first and last name etc.

Fig 9. User created folders and files as per preferences

FILE TYPE WHICH CAN BE SUBMITTED

- MS Word,
- Word XML,
- WordPerfect,
- Postscript,
- PDF,
- HTML, RTF, Open Office (ODT) and plain text

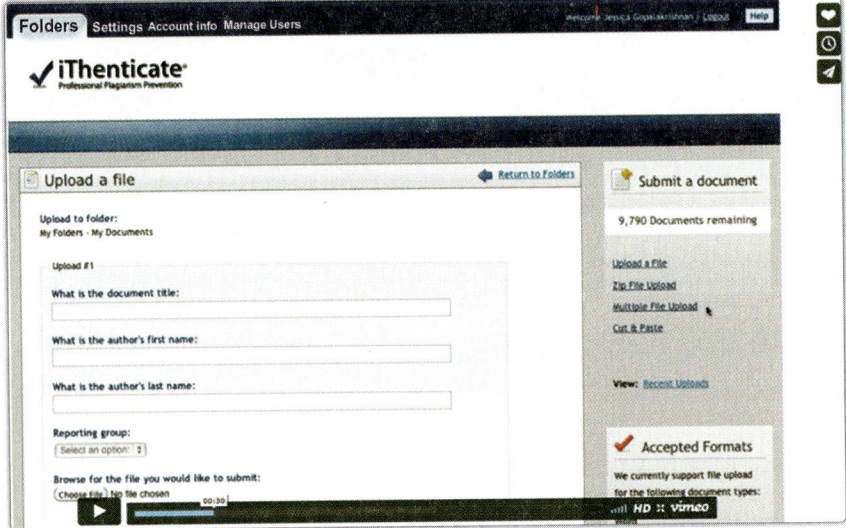

Fig 10. Window for uploading the file

Once the document is submitted for the similarity check, a report is generated on the dashboard of the user, which he can easily open and seen the amount of plagiarism, his/her document suffers with.

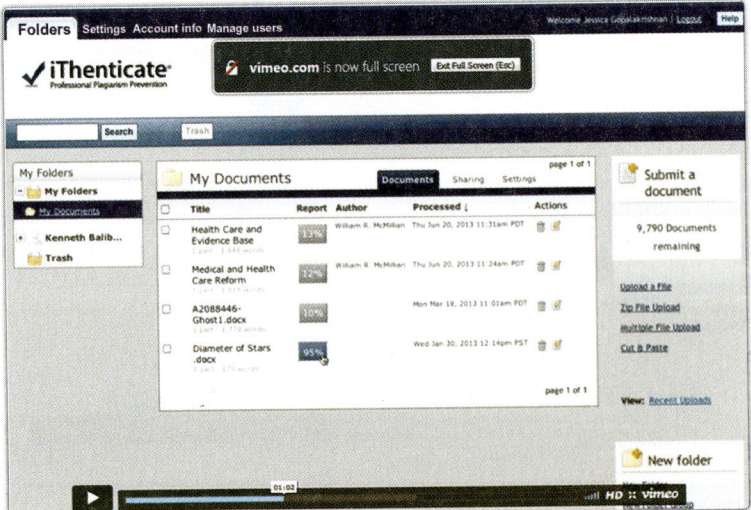

Fig 11. Window to view the overall report of the document submitted for scan

To view the complete report of the document, one needs to click on the percentage text reflected as similar to some other sources. This will lead the user to see all such sources where the text resembles similar to various other sources. The text reflected as similar to other sources is mostly highlighted in different colours. Each individual colour represents the each individual source where from the text has been taken. One can directly link the sources reflected itself in the document.

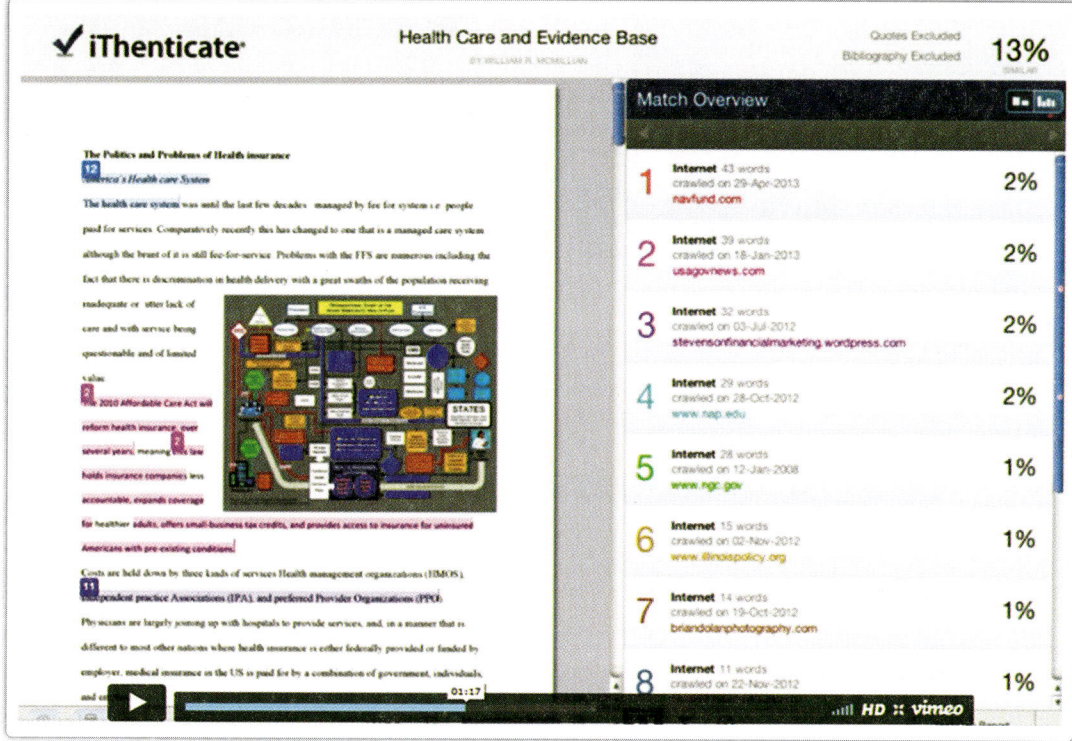

Fig 12. Window to view the exact text shown as matching with other sources

TEXT MATCHING

- iThenticate checks against the following databases :
- **CrossCheck:** It is an initiative of the publisher association CrossRef, CrossCheck is a database populated with content (journal articles, conference proceedings and books) provided by more than 300 publishers.
- Internet archive iParadigms's proprietary Internet crawler is archived back nearly a decade.
- Aggregators, databases, content providers Over 90 million online and offline subscription content and research titles from leading aggregators, databases and content providers
- ProQuest Dissertations and Theses (PQDT) is a research database that includes theses and dissertations from more than 1,000 North American and European graduate schools

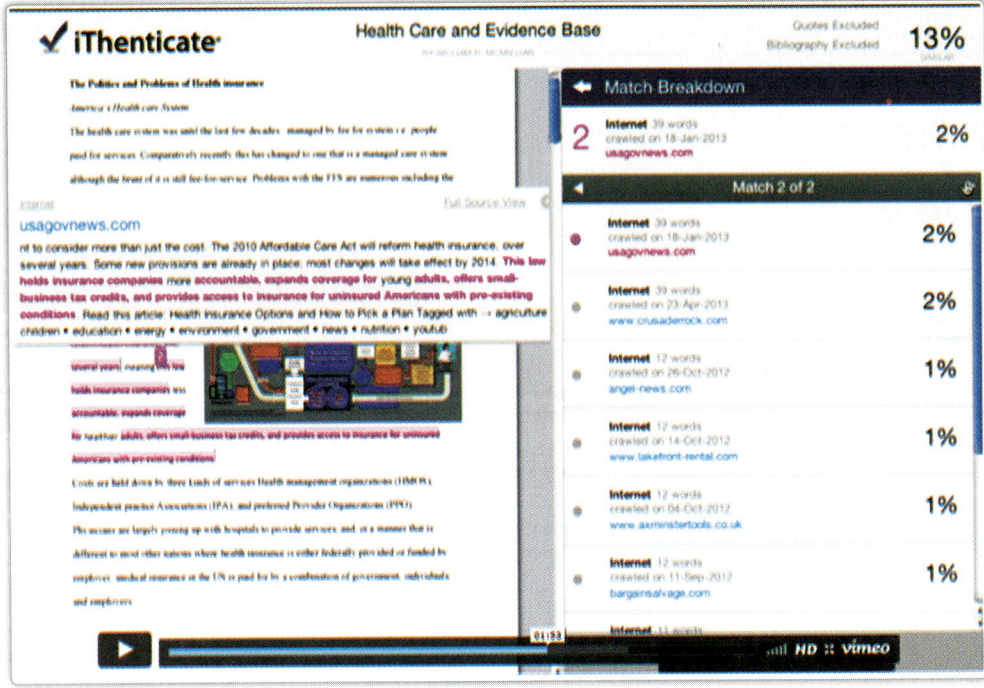

Fig 13. Window to view the sources with which document matches

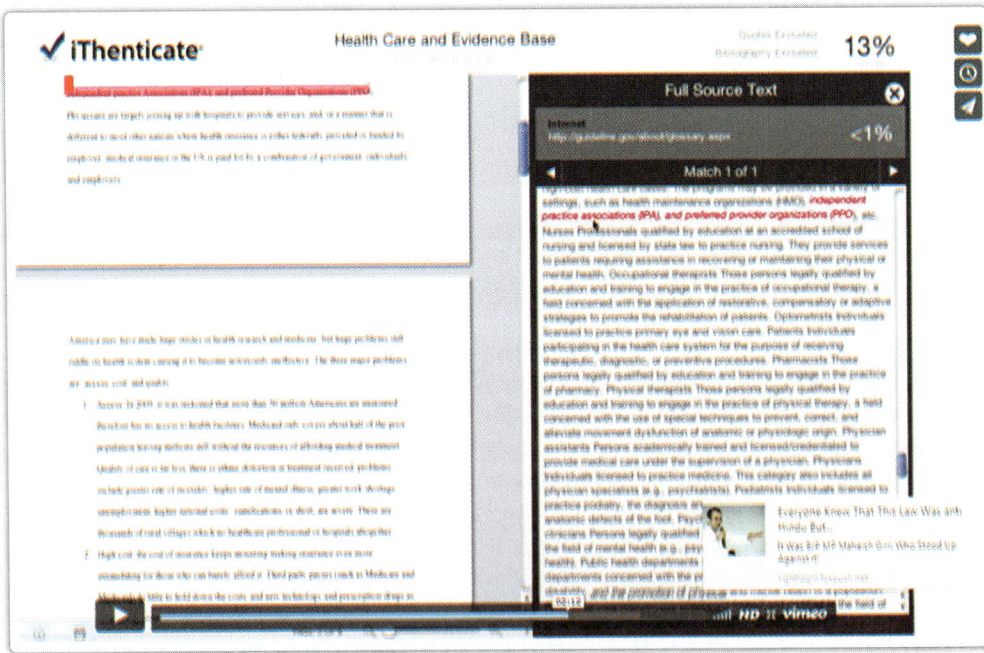

Fig 14. Window to view the exact text taken from the original source

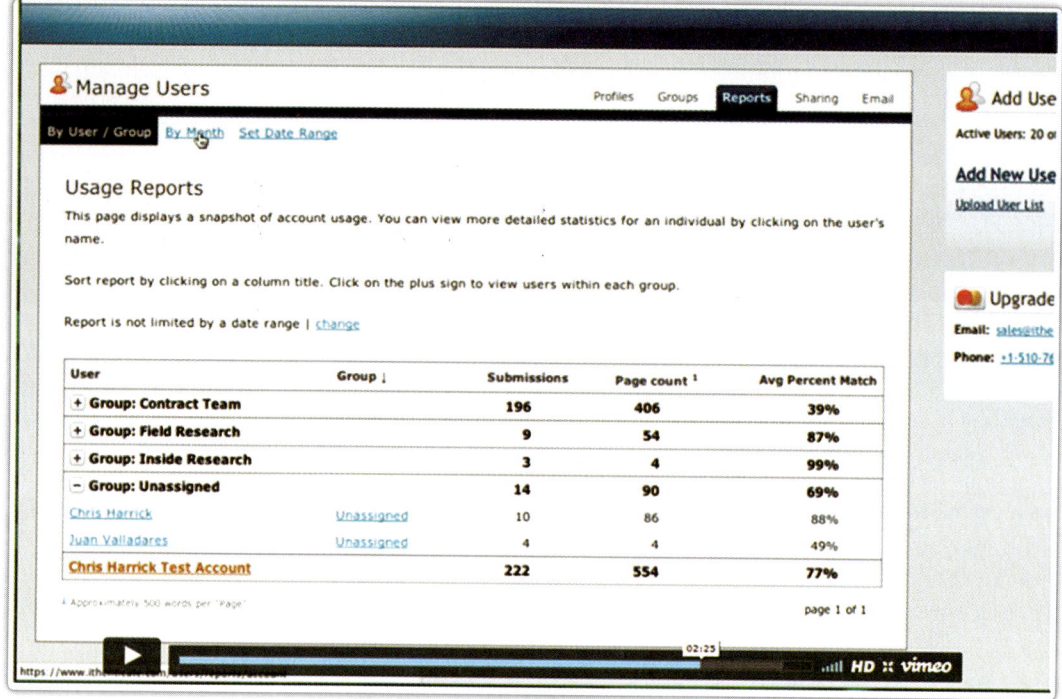

Fig 15. Overall report of the documents submitted by the user for scan

After going through the complete report the researcher can take out necessary corrections in the final document or modify it keeping in view the scope of his/her study and the need to improve upon.

HELP ON ITHENTICATE

- Documentation and training materials are available on the iThenticate website: http://www.ithenticate. com/resources/customer-training

URKUND

Of late University Grants Commission (UGC), New Delhi through its Inter-University Centre, INFLIBNET, Ahmedabad has made available Urkund software to universities, recognized by UGC under section 2 (F) and 12 (B) of the UGC act of 1956.

Like other two software discussed earlier, Urkund is equally a popular software used for checking documents online for similar content, if any, what we commonly know as plagiarism.

The URKUND software is owned and developed by Prio Info Sweden. Since the development of software in the year 2000, URKUND has gained popularity all across the world, especially among the academic world. The software enjoys a fair amount of popularity all across the world, especially in the regions like, Europe, USA, Asia, and the Middle East. The software is popularly used by the researchers, students and the teachers all across the world. The software not only helps in detecting the plagiarism of the scholarly and other created content, but also deters the information creating community to desist from every sort of academic misconduct. The software is equally used by the organizations, which are information sensitive.

The use of plagiarism detection software gained popularity with the unprecedented growth of information on the world wide web. The plagiarism detection software can be effectively used for checking the documents for similarity against the documents available online. Given the fact, the content published in the electronic format can be easily copied, cut and pasted, hence emerged as one of the easiest means of indulging into academic misconduct. The plagiarism detection software are also used effectively by the academia all across the world to reduce the instances of plagiarism. Most of the researchers these days scan their preprints agsint the plagiarism detection software for similarity check and wherever needed the instances of plagiarism can be reduced to a considerable level before the final document is published.

URKUND is very user friendly and easy to use software for similarity check. The students, scholars and the teachers of the institutions which subscribe the software can easily register on the URKUND at its portal, accessible at www.urkund.com

The software works on the pattern by creating an institutional administrator to handle the operation of software by assigning an analysis address. The users have to simply register themselves on the software portal and have to provide the valid analysis address, while submitting the document for similarity check. There is need to understand that user who register themselves on the software portal can only submit the documents for similarity check, but cannot access the report. The report generated is generally mailed to the admin at his email address, which is linked to the analysis address of the institution. The admin thereafter can forward the detailed report to the user at his/her email address, which is readily provided with a report, as who has submitted this particular document for similarity check.

The user can register themselves on the URKUND portal, which is accessible at www.urkund.com

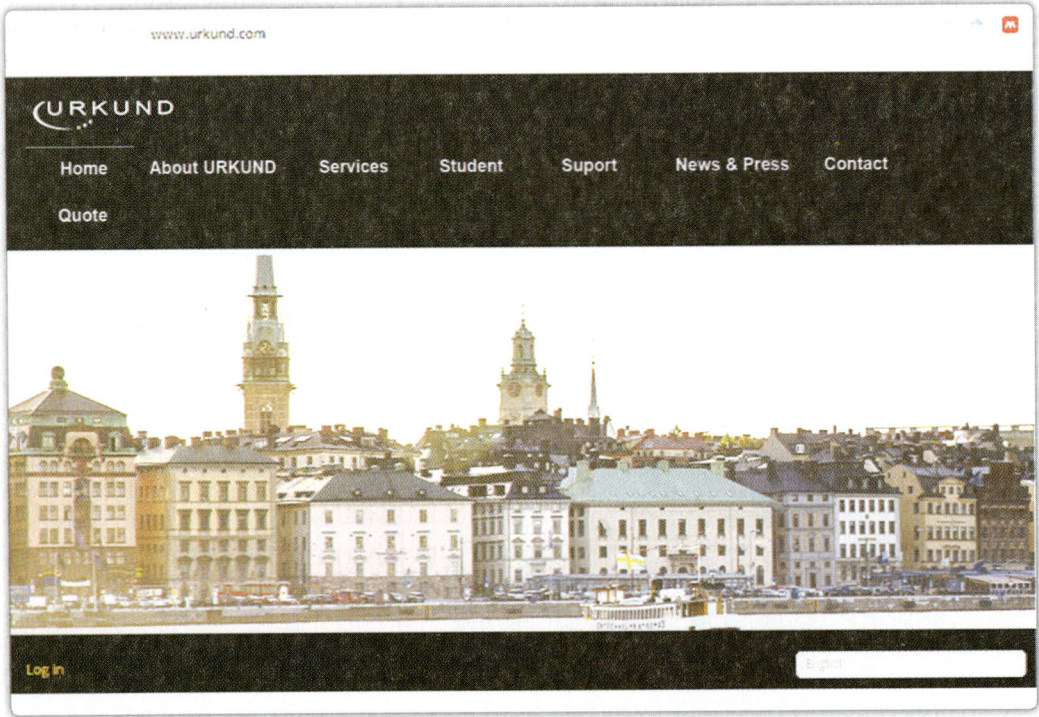

Fig 16. Main window of the Urkund software

Users can follow screen shots as their guiding steps to register with Urkund and use it effectively for their academic purposes.

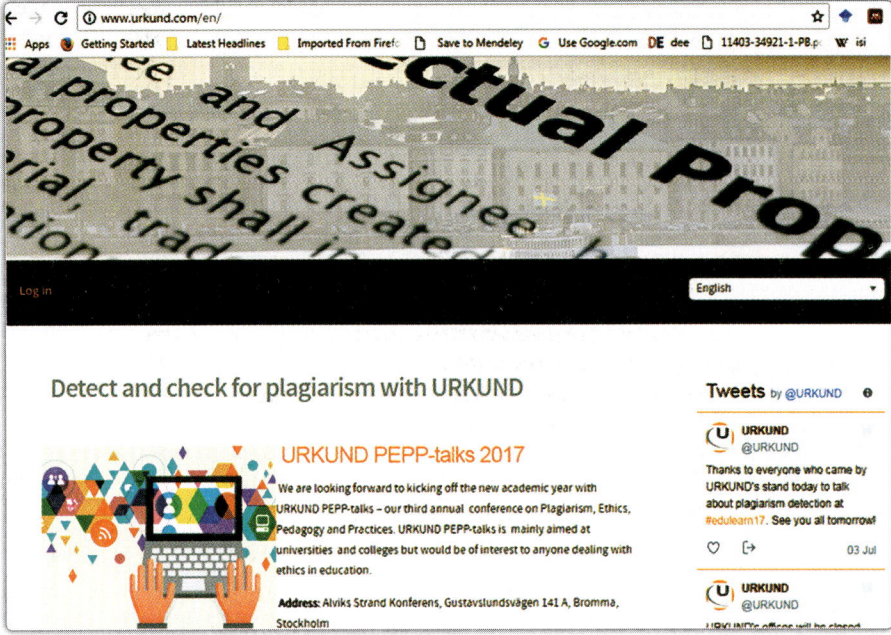

Fig 17. Users login window

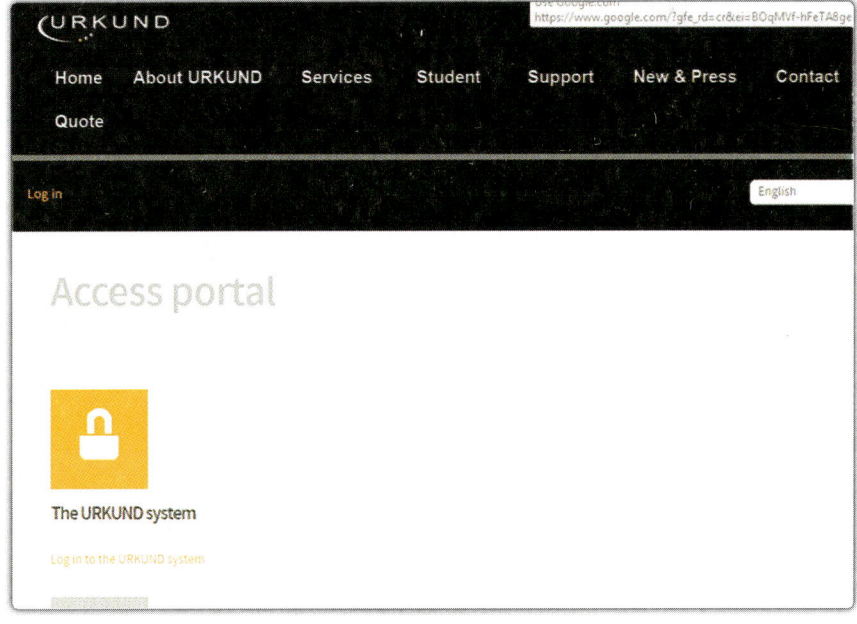

Fig 18. User login window-II

After registering, the users can simply login by using the username and password, as shown in the below mentioned figure so as to reach to their dashboard.

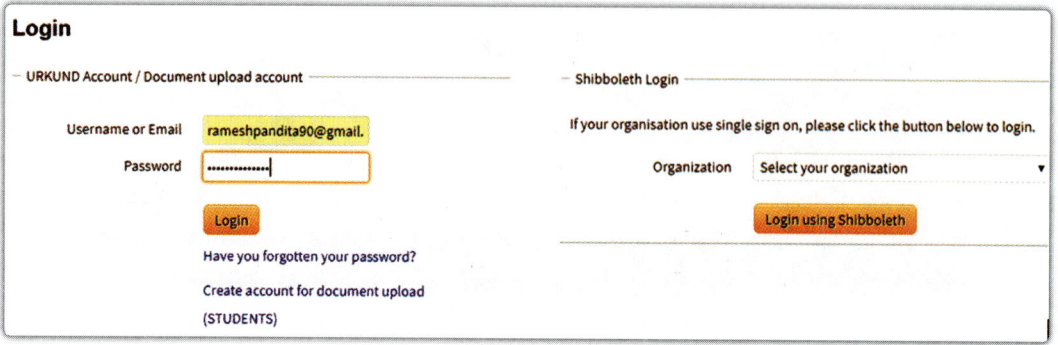

Fig 19. Window to fill the login credentials

Analysis Address : rameshpandita90.bgsbu@analysis.urkund.com

8%	D29599285 Review IJIDT_0750 (1).doc	110 KB	5523 word(s)	Ramesh Pandita	7/11/2017 8:36 AM
1%	D29559839 Thesis_fi_shaa_allah_-_Copy (1).docx	462 KB	47404 word(s)	Ramesh Pandita	7/6/2017 5:46 AM
0%	D29559838 Abstarct (1).docx	38 KB	3952 word(s)	Ramesh Pandita	7/6/2017 5:46 AM
2%	D29527463 Thesis_fi_shaa_allah_-_Copy.docx	483 KB	50016 word(s)	Ramesh Pandita	6/30/2017 10:03 AM
17%	D29527462 Abstarct.docx	38 KB	4060 word(s)	Ramesh Pandita	6/30/2017 10:03 AM
0%	D29520366 rp modified.docx	91 KB	1273 word(s)	ed.gowhar@gmail.com	6/29/2017 10:35 AM
10%	D29466789 thesis_final (2).docx	466 KB	65995 word(s)	Ramesh Pandita	6/22/2017 9:30 AM
3%	D29461843 AERJ-2017-033.doc	115 KB	5396 word(s)	Ramesh Pandita	6/21/2017 4:54 PM
40%	D29459559 rp.docx	87 KB	1700 word(s)	ed.gowhar@gmail.com	6/21/2017 2:13 PM
11%	D29447414 thesis_final.docx	466 KB	65902 word(s)	Ramesh Pandita	6/20/2017 10:34 AM
29%	D29420171 11406-31968-2-RV.doc	378 KB	3580 word(s)	Ramesh Pandita	6/17/2017 7:02 AM
0%	D29391793 chapter 1 o.docx	84 KB	7685 word(s)	inshamn@gmail.com	6/15/2017 11:39 AM
1%	D29391699 chapter 2.docx	821 KB	10738 word(s)	inshamn@gmail.com	6/15/2017 11:33 AM
2%	D29391698 CHAPTER 4.docx	706 KB	20399 word(s)	inshamn@gmail.com	6/15/2017 11:33 AM
1%	D29391697 chapter 3.docx	59 KB	7732 word(s)	inshamn@gmail.com	6/15/2017 11:33 AM
5%	D29388016 5_summary and conclusions.doc	143 KB	2768 word(s)	Ramesh Pandita	6/15/2017 6:04 AM
1%	D29388013 4_Chapter_4_final.doc	13 MB	7743 word(s)	Ramesh Pandita	6/15/2017 6:04 AM

Fig 20. Dashboard of the user

Before uploading the document the software will prompt the user for analysis address. So every time the user uploads a document for the similar check should keep the analysis address handy with them. Generally the software administrators circulate the analysis address among students, scholars and the faculty member of their respective institution, so as to facilitate the uploading of documents among the user community.

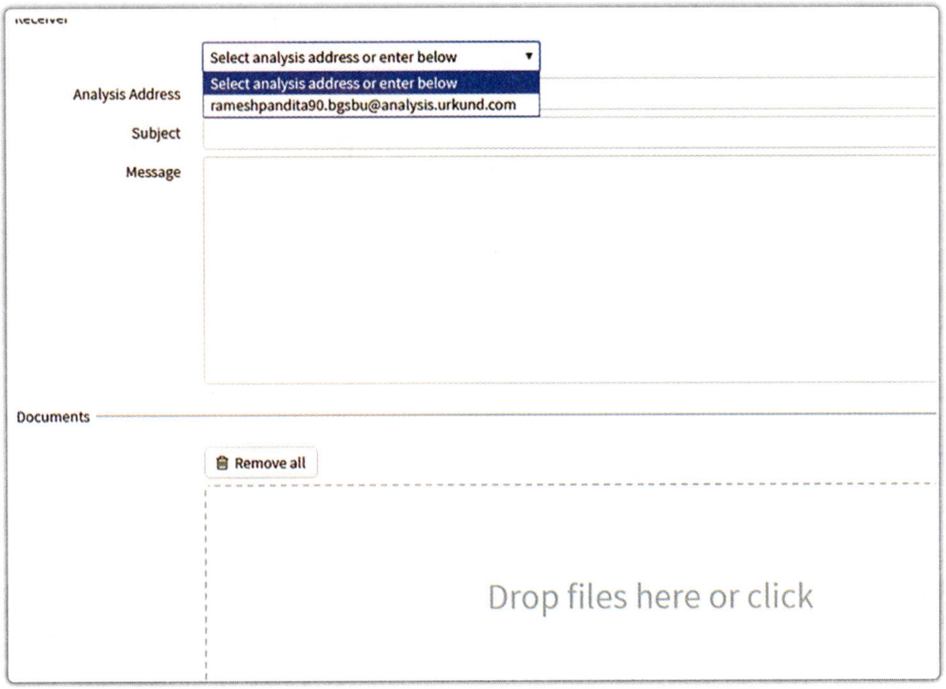

Receiver

Analysis Address | Select analysis address or enter below ▼ |
| Select analysis address or enter below |
| rameshpandita90.bgsbu@analysis.urkund.com |

Subject

Message

Documents

🗑 Remove all

Drop files here or click

Fig 21. Window to fill the analysis address provided by the service provider

The moment valid analysis address is provided in the analysis address bar, the address bar will reflect the institutional name along with the analysis address.

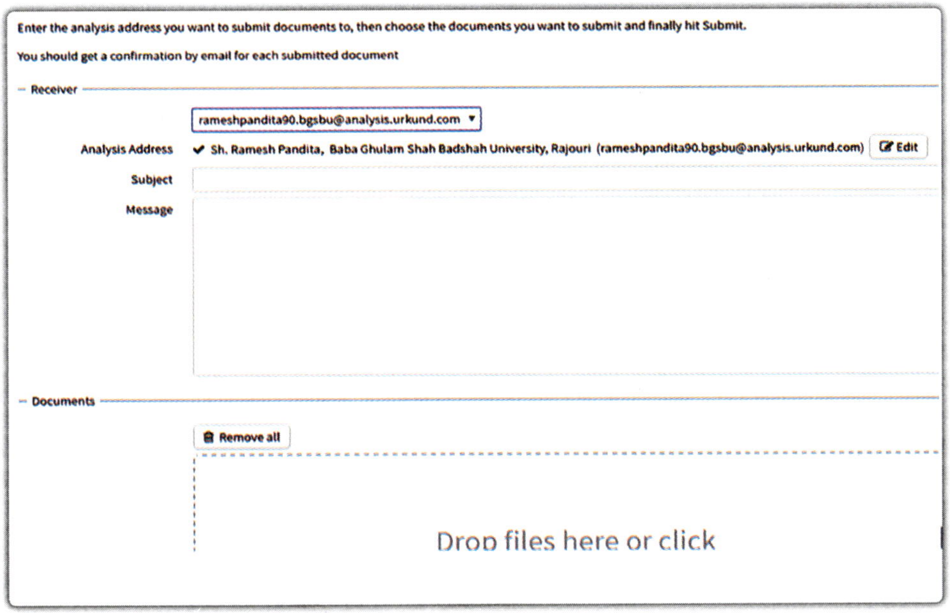

Enter the analysis address you want to submit documents to, then choose the documents you want to submit and finally hit Submit.

You should get a confirmation by email for each submitted document

— Receiver

| rameshpandita90.bgsbu@analysis.urkund.com ▼ |

Analysis Address ✔ Sh. Ramesh Pandita, Baba Ghulam Shah Badshah University, Rajouri (rameshpandita90.bgsbu@analysis.urkund.com) ✔ Edit

Subject

Message

— Documents

🗑 Remove all

Drop files here or click

Fig 22. Window to fill the analysis address-II

Users can submit the documents for similarity check either by dragging and dropping the files on the space provided on the window or by clicking on the space and then selecting the file location on the computer, as is done while attaching the files in an email.

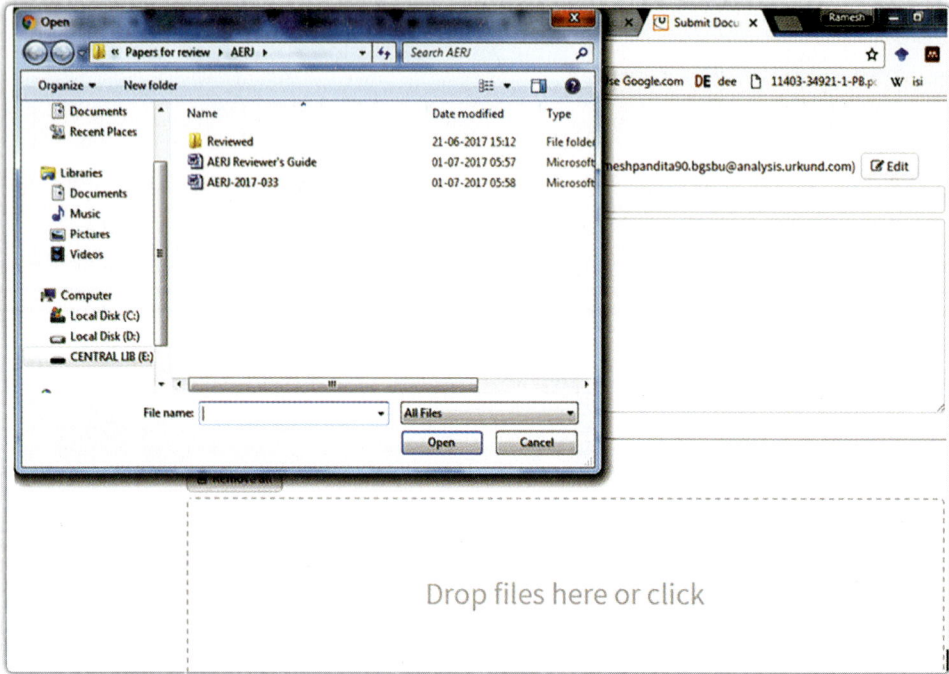

Fig 23. Window to submit the document from desktop or other attached device

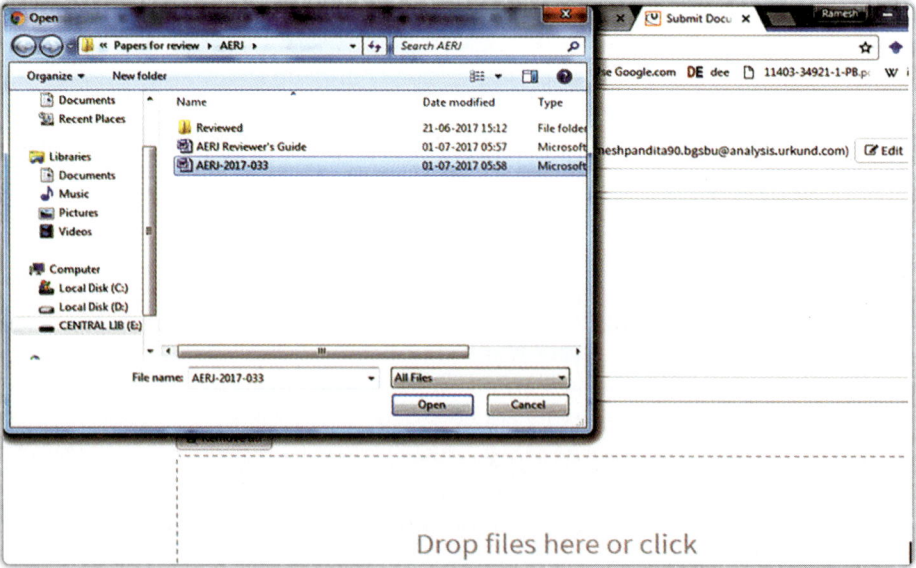

Fig 24. Window to submit the document from desktop or other attached device-II

Once the file is uploaded, the file will be reflected on the space provided on the upload window along with complete information like, the file name, file size, file format etc. In case the file is not the one which the user had intended for similarity check, the user can simply click on the remove button, the file will be deleted. The procedure can be repeated for uploading the correct file.

Fig 25. Window highlighting the document uploaded for submission

After the desired file has been uploaded, then click the submit button provided at the bottom of the document upload window.

Fig 26. Window to submit the document by clicking the submit button

Once the submit button is clicked, the window will show the status bar of the document submission in green colour and so will the submitter of the document receive the confirmation notification of the document submission.

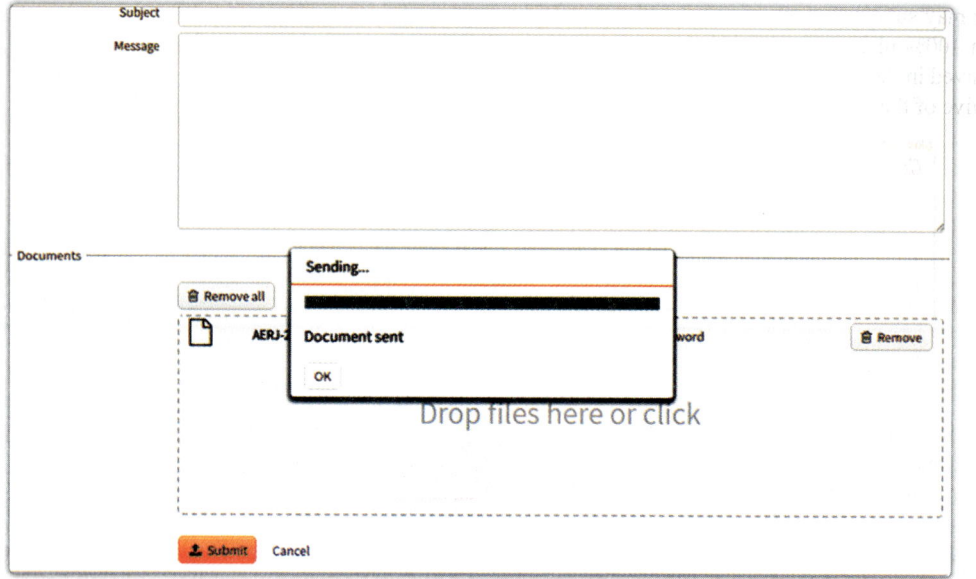

Fig 27. Document submission status window

The user can check the submitted document on his/her dashboard along with unique document ID assigned to each individual document on its submission.

			Document			
			D29606412 AERJ-2017-033.doc	117 KB	5397 word(s)	Ramesh Pandita	7/11/2017 4:17 PM
		8%	D29599285 Review IJIDT_0750 (1).doc	110 KB	5523 word(s)	Ramesh Pandita	7/11/2017 8:36 AM
		1%	D29559839 Thesis_fi_shaa_allah_-_Copy (1).docx	462 KB	47404 word(s)	Ramesh Pandita	7/6/2017 5:46 AM
		0%	D29559838 Abstarct (1).docx	38 KB	3952 word(s)	Ramesh Pandita	7/6/2017 5:46 AM
		2%	D29527463 Thesis_fi_shaa_allah_-_Copy.docx	483 KB	50016 word(s)	Ramesh Pandita	6/30/2017 10:03 AM
		17%	D29527462 Abstarct.docx	38 KB	4060 word(s)	Ramesh Pandita	6/30/2017 10:03 AM
		0%	D29520386 rp modified.docx	91 KB	1273 word(s)	ed.gowhar@gmail.com	6/29/2017 10:35 AM
		10%	D29486789 thesis_final (2).docx	466 KB	65995 word(s)	Ramesh Pandita	6/22/2017 9:30 AM
		3%	D29461843 AERJ-2017-033.doc	115 KB	5396 word(s)	Ramesh Pandita	6/21/2017 4:54 PM
		40%	D29459559 rp.docx	87 KB	1700 word(s)	ed.gowhar@gmail.com	6/21/2017 2:13 PM
		11%	D29447414 thesis_final.docx	466 KB	65902 word(s)	Ramesh Pandita	6/20/2017 10:34 AM
		29%	D29420171 11406-31968-2-RV.doc	378 KB	3580 word(s)	Ramesh Pandita	6/17/2017 7:02 AM
		0%	D29391793 chapter 1 o.docx	84 KB	7685 word(s)	inshamn@gmail.com	6/15/2017 11:39 AM
		1%	D29391699 chapter 2.docx	821 KB	10738 word(s)	inshamn@gmail.com	6/15/2017 11:33 AM
		2%	D29391698 CHAPTER 4.docx	706 KB	20399 word(s)	inshamn@gmail.com	6/15/2017 11:33 AM
		1%	D29391697 chapter 3.docx	59 KB	7732 word(s)	inshamn@gmail.com	6/15/2017 11:33 AM
		5%	D29388016 5_summary and conclusions.doc	143 KB	2768 word(s)	Ramesh Pandita	6/15/2017 6:04 AM

Fig 28. Dashboard of the user reflecting the documents submitted earlier and their overall plagiarism percentage

Confirmation receipt mail will be sent to your email, along with complete document details along with a couple of links, which may prompt the user to archive the document in the Urkund. Here the users have to make sure to not to archive the document in the Urkund, unless required. As by doing so, the next time the user may submit the same document for similarity check, the report may reflect as the document suffering with 100% plagiarism. This happens because the same document will be scanned against the documents archived in the Urkund. So, one has to be very careful in this regard before archiving his/her document in the archive of the software.

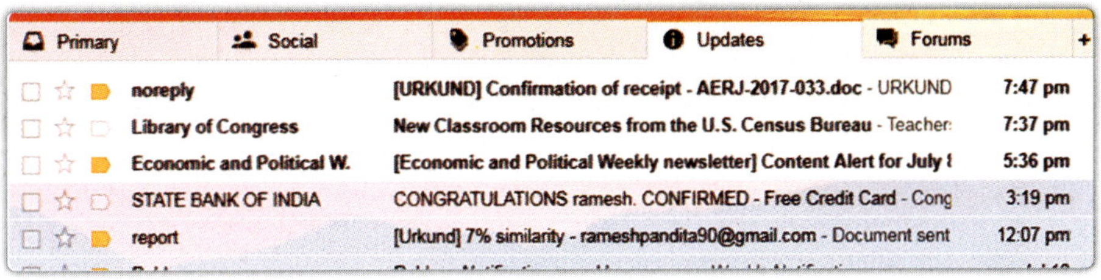

Fig 29. Receipt of document submission forwarded to users email id

This will be followed by the final report, which is always mailed to the institutional admin. A clientele who wants his/her document to be checked for similarity of content can itself upload the document, but cannot view the report directly. The report is first forwarded to the admin in whose name the analysis address of the institution is registered and thereafter same can be forwarded to the concerned person.

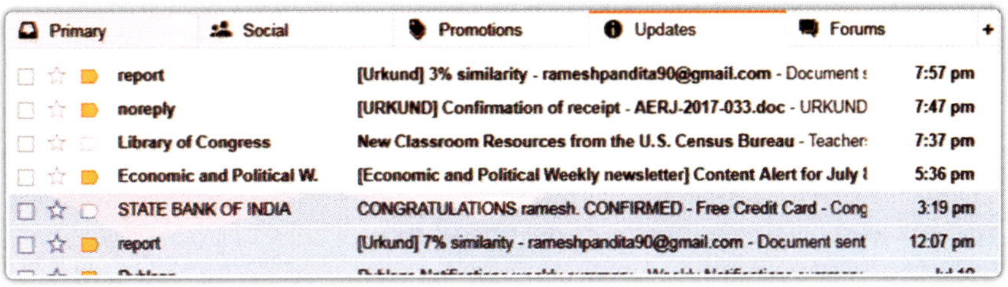

Fig 30. Report of plagiarism forwarded to the user in his/her email id

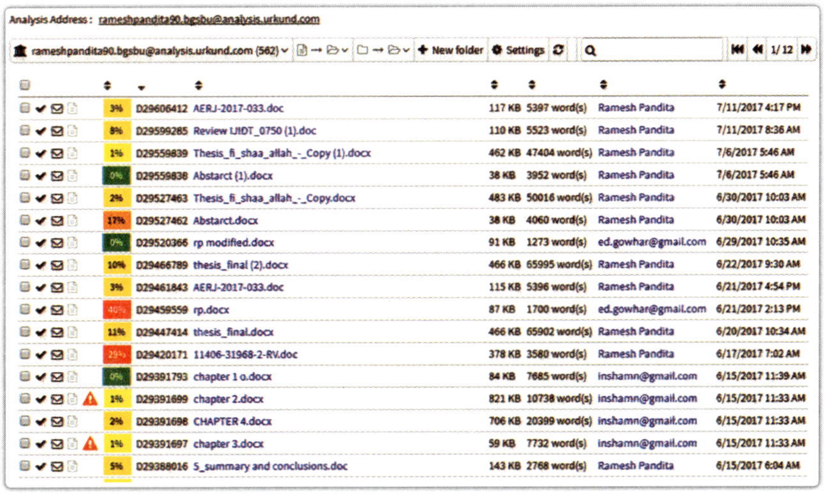

Fig 31. Window reflecting the overall plagiarism percentage of the document

One can view the report on screen by following the analysis link or even can download it by simply exporting the document.

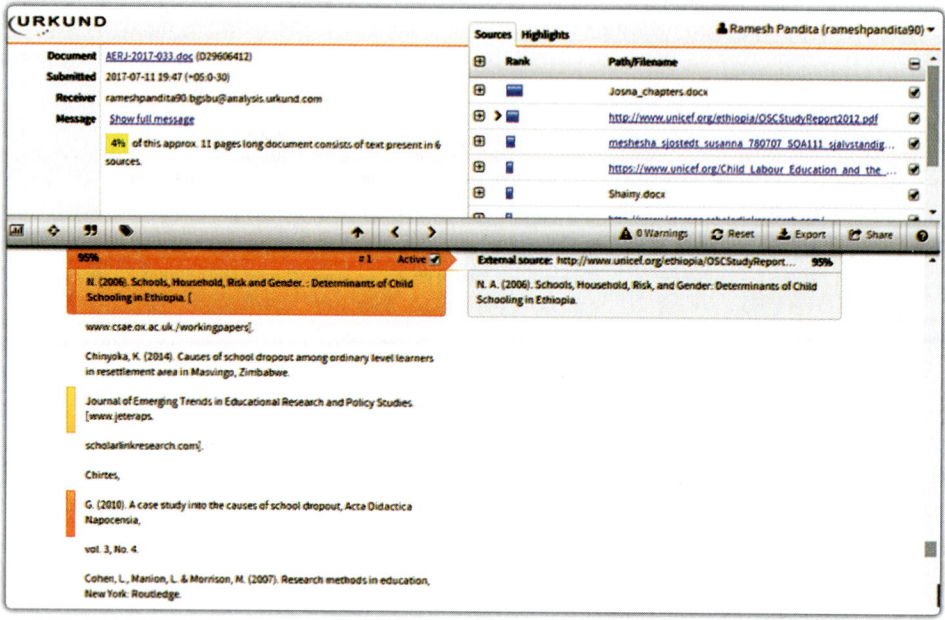

Fig 32. Window to view the report

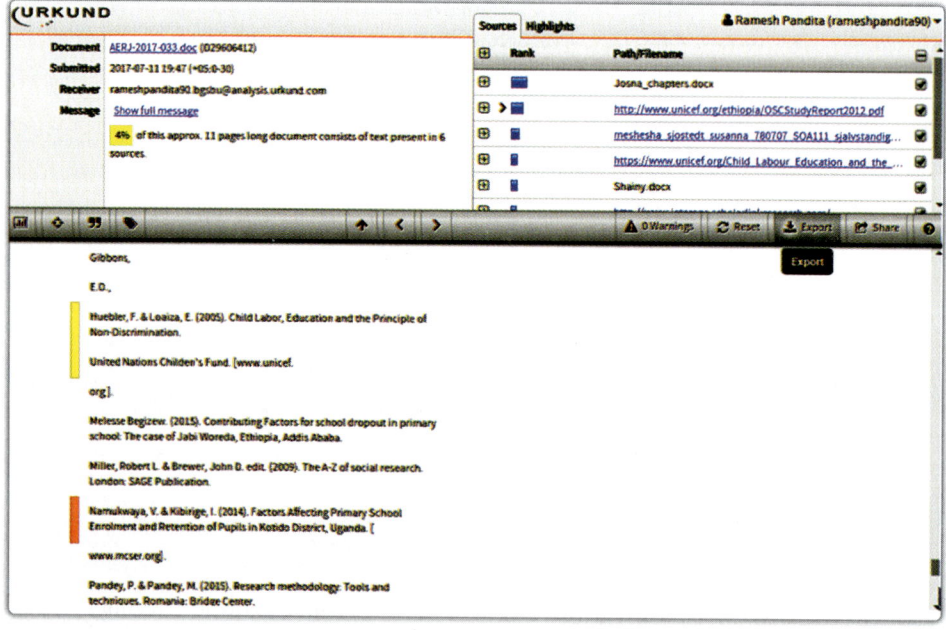

Fig 33. Window to view the report-II

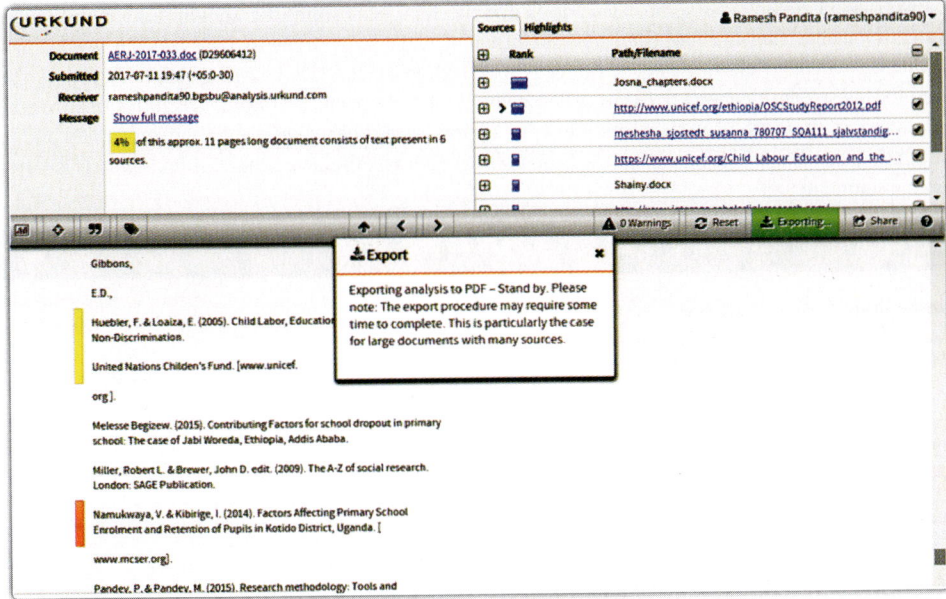

Fig 34. Window to download the report

On the very first page of the downloaded report reflects the significance level of the similarity along with other complete document details. The report also highlights the number of instances where text is found similar to other sources.

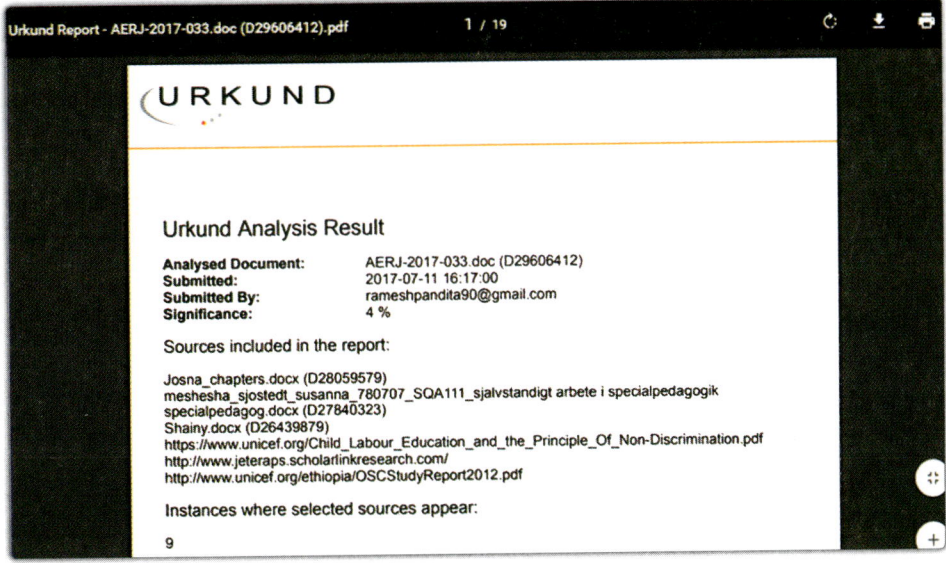

Fig 35. Downloaded report in pdf format

In the downloaded report, users can easily look for the text which may be shown similar to some other text, generally wherefrom the text may have been taken as it is. In order to facilitate the users to review the document, two parallel windows of the text are displayed. The left window generally shows the text submitted for the similarity check or the host document which has been submitted for similarity check and the right side window shows the source document wherefrom the text has most possibly been copied and pasted to a certain degree. Each highlighted part of the text even shows the percentage of text, which is similar to the source document.

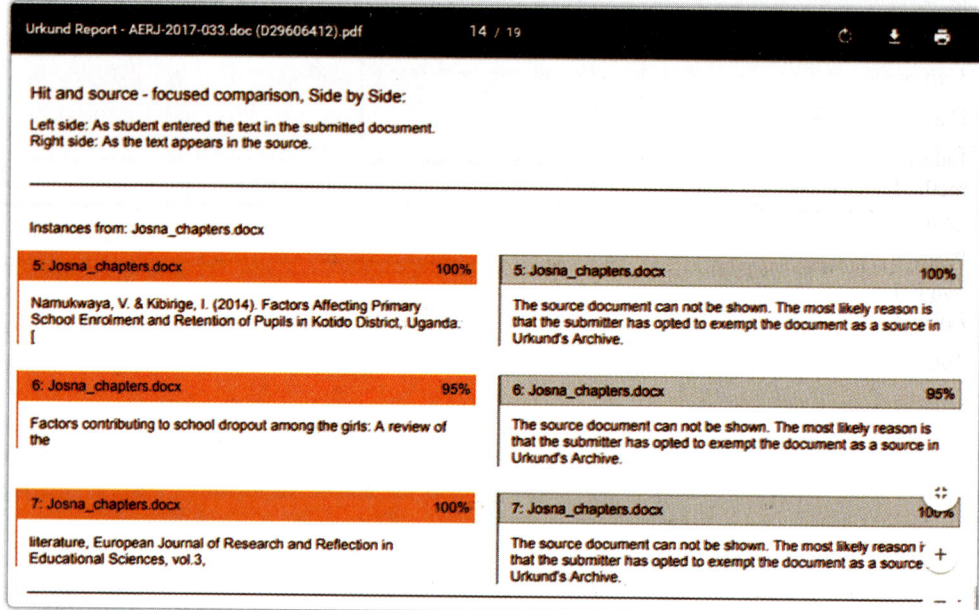

Fig 36. Report showing the matching content

On the whole Urkund is a very user friendly software which can be easily used for similarity check by institutions, for students, scholars and the faculty members. The researcher can benefit in many ways, they can easily do away with those parts of the text, which may fall under plagiarism. The researchers can also do some paraphrasing and prepare the text with a whole new idea and objective in the light of research works already carried out. It is always advisable that academic institutions should subscribe to any of the plagiarism detection software's of their choice for the benefit of researchers in general and the institution as a whole in particular.

CONCLUSION

The plagiarism detection tools viz., software help in checking the documents against those documents for similarity check, which are available on the web, hosted on different servers across the globe. Ultimately, it is the research experts who are left with the responsibility to decide as whether the content detected similar to other documents or sources amounts to plagiarism or not. These plagiarism detection tools help a great deal in avoiding plagiarism and in refining the content. The market of plagiarism detection tools is quite promising as most of the academic and research institutions are subscribing the services of plagiarism detection software and so are these tools being put to use by publishers to detect the level of plagiarism before going ahead with the final publication of content in a journal or a book. The students and the scholars are equally benefited of these tools in refining their content before submitting their projects, research thesis and other reports to their respective institutions.

BIBLIOGRAPHY

1. Blum, S. D. (2011). *My word!: Plagiarism and college culture*. Cornell University Press.

2. Carroll, J., & Oxford Centre for Staff Development. (2007). *A handbook for deterring plagiarism in higher education* (Vol. 2). Oxford: Oxford Centre for Staff and Learning Development.

3. Cebrian, M., Alfonseca, M., & Ortega, A. (2009). Towards the validation of plagiarism detection tools by means of grammar evolution. *IEEE Transactions on Evolutionary Computation, 13*(3), 477-485.

4. Hage, J., Rademaker, P., & van Vugt, N. (2010). A comparison of plagiarism detection tools. *Utrecht University. Utrecht, The Netherlands*, 28.

5. iThenticate (n.d). Plagiarism detectin software. Retrieved from http://www.ithenticate.com/

6. Lukashenko, R., Graudina, V., & Grundspenkis, J. (2007, June). Computer-based plagiarism detection methods and tools: an overview. In *Proceedings of the 2007 international conference on Computer systems and technologies* (p. 40). ACM.

7. Potthast, M., Stein, B., Barrón-Cedeño, A., & Rosso, P. (2010, August). An evaluation framework for plagiarism detection. In *Proceedings of the 23rd international conference on computational linguistics: Posters* (pp. 997-1005). Association for Computational Linguistics.

8. Turnitin (n.d). Plagiarim detection software. Retrieved from https://turnitin.com/gateway/index.html

9. Urkund (n.d). Plagiarism detection software. Retrieved from http://www.urkund.com/

Chapter 4

Research Writing

INTRODUCTION

Writing a research paper requires a lot of efforts, focused mind, patience and what not, but apart from these, there are some basic tips and techniques with the help of which a researcher can easily fine tune his/her research article. Accordingly, the chapter deals with some very common techniques, which a researcher can employee while writing a research article. With the help of these techniques one can produce a good quality research articles with minimum efforts and far greater ease. In the undergoing deliberations, some important components have been discussed, which can prove helpful and handy in writing research papers/dissertations/thesis, etc. There is always need to structure and coordinate ones research writing in the most convenient way, given the conditioning and the suitability of the research report or research article writing. Attention has been invited toward undertaking a substantive, sophisticated and thorough literature review which more or less is a precondition to undertake a substantive, sophisticated and thorough research. Emphasis has also been laid on the need and importance of attributions to be made both within and at the end of the research work in the form of references, bibliography and acknowledgements along with the need to adhere to different referencing and citation styles.

VARIOUS ASPECTS OF RESEARCH WRITING

Method

This is to undertake any activity or to execute any act, one has to choose a certain method of doing it. Accordingly, when we undertake any research activity we follow a certain path what we generally call as a research method. A research activity is always supposed to be undertaken by adhering to certain research principles and norms generally laid down in each and every kind of science. Research results cannot be held as valid unless the research methods adopted by the researcher to undertake the research activity are not regarded as valid and acceptable.

Findings/Results

Most of the research results are presented in empirical form. It is always important for a researcher to present his/her research findings in such a manner, which can be easily understandable, which can be easily verified, which can be easily tested and tried. The researcher should present his/her findings, keeping in view the end users and should not merely end up with a mindset that findings have no practical purposes.

Relevance and Impact of the Study

As discussed above, it is always the end user or the actual beneficiary who is to be kept in mind while undertaking any research activity. There is no denial of the fact that most of the research work undergoing across the world is generally aimed at the welfare and the betterment of humans, but still we may come across those research activities, which may not turn out to be need based and may show no relevance to the human needs. Although, research is always supposed to be pursued in the direction, a researcher shows interest in, but still it is always desirable that most of the research activities and actions should be focused on the immediate needs of the end users.

Word Limit

Research writing is both art and skill. A researcher has to be always careful and vigilant, while writing a research report. A research report should be always brief, precise and to the point. A waywardly written research report can bog down the whole purpose of research and may even turn the valuable findings into meaningless observations. A lengthy research report can disinterest the readers, hence may defeat the purpose of whole research. It is always advisable to keep your research report as brief as possible. Although, there is no hard and fast rule about the word limit, but surely anything between 2,000 and 4,000 words is the general norm of submitting reports. This trend of research report writing again varies depending on the subject discipline. It has been observed that in natural and pure sciences, research reports are brief, while as in social sciences, research articles are generally lengthy in nature.

MAIN COMPONENTS

A research paper has different components, which are somewhere the perquisites to establish the validity of research paper. Some of the key components associated with the layout of a research paper are as under:

Title

Title of a research paper forms the very basis of any research writing. Title of a research article acts as a bridge between the content and the information seeker. It is always the very title of any book, research article or any other popular article, which catches the attention of the reader then only the reader decides whether to go through the entire article or not. For any researcher, it is always imperative to be very careful in choosing the title of a research article. A title should always be logical, brief and rigorous enough to reflect the essence of the research article or publication. Titles are also sometimes continued in the form of sub-titles depending upon the amount of essence a researcher wants to put in a title to attract his/her peers toward his/her research findings. Accordingly, considerable difference is found in the titles of the research articles in pure sciences, social sciences and arts and humanities, etc.

Author and His Organization

To produce research findings or research report is always questioned for its credibility and authority. Unless a research article does not reflect the issuing authority be it the person, institution or an organization, which may have undertaken the research, the research reports hold no validity. The research report or findings are never accepted, unless the name affiliation with an institution, his/her status, etc. of the researcher are not mentioned. All this information should always be reflected in the article, firstly to claim the right of the intellectual property and secondly to make it more trustworthy, so that people may not question the authority or credibility of the research findings.

Abstract

It is being always said that titles should deceptive, as we may have a book or article which may have a very catchy title, but may not necessarily turn out to be a catchy book or article as a whole. While writing books, apart from reflecting contents we also write a preface of the book. This preface reflects the complete idea about the whole nature of the book and the type of content and topics a particular document deals with. Similarly, abstract deals with the type of research undertaken in a particular article and may also reflect some findings to give a clear idea about the research paper. So, it is always advisable to reflect the crux of the study at the beginning in the form of an abstract. An abstract is a summary of the whole research article, presented mostly between 100 and 200 words. After going through the abstract a reader is to decide whether he/she should go through the whole research article or not.

- The abstract is not an introduction
- It is a description of the whole paper in a concise manner, highlighting important points.
- Self-contained
- No references

Abstracts are of two types, one is informative and other is indicative. An informative abstract is that abstract, which also reflects a good percentage of research results in it. Readers most of the time, after going through these abstracts need not to go through the entire document or the paper, as their information purpose may get itself fulfilled by going through the abstract. Contrary to it, an indicative abstract simply reflects the central idea about the publication and it is the reader who, after going through the abstract has to decide whether he/she has to go through the entire paper or not.

Keywords

Keywords are considered as an important and integral part of both academic and scientific writing. Keywords represent the broader subject area in a single word. Most of the keywords assigned to a research paper are sequential in nature and reflect a sort of events defined, discussed and deliberated in an article. In the present day world of information and technology, keywords play an important part in the search strategy of any information or research article. The information seeker has to be very precise with the keywords to materialize the prized catch of the desired piece of information. Truncation forms an important tool for information search strategy and helps a researcher to confine the search to limited words or terms and thus helps in retrieving the desired piece of information. Apart from above, some of the other key features of keywords are:

- Keywords should appropriately describe the essence or subject matter
- Facilitates online searches
- Choose the keywords, which are relevant, representative and broad based in use so as to be searched quickly on the Internet a paper exactly like yours
- Try to choose the keywords different from that of the title.

Introduction

In general terms, introduction is about passing on basic information about a person, place or a thing between the transmitter and the recipient. Introduction between two persons is simply about exchanging basic information like name, profession, place where one lives, work and other social standings, etc. about each other, so as to gain familiarity and knowhow about each other. However, when we talk about the introductory part of any research article or any academic paper, we simply reflect the basic information about the research article, to give an idea to the readers about the complete research paper. In the research article, researchers simply reflect the research problem, need, purpose and importance of the research article. Apart from these, the introductory part of any research, writing talks about the following aspects of research paper:

- Nature of the problem
- Essence of the state of the art in the domain of the paper
- Goals/aims of the paper, its relevance for pushing forward the state of the art
- Methods used to solve the problem
- The structure of the paper, describing briefly the successive sections.

Body of The Research Paper

Body of a research paper generally refers to all the sections and the subsections, which together form the whole of the paper. This includes the title of the research article to the references cited at the end of an article. Some of the key concepts, which constitute the body of the research article apart from abstract and introduction include:

- Background information on the research problem
- Background information on the research area, if it is region-specific, etc.
- Objectives/hypothesis etc.
- Scope/methods/limitations of the study
- Results/data analysis, etc.
- Discussion, suggestions, recommendations, finding, conclusion, etc.

All these body parts of the research article talk and discuss about, as how the study is designed to answer the problem or research question.

Methodology Adopted

It is always desirable to discuss about the research methods adopted or followed, while following a particular research procedure. Research methods, in fact, act as a backbone for research which is important to validate the research findings. A research method discussed in any research report corroborates the findings and also helps in verifying the research results. Research methods are also necessary from the point that these methods substantiate the procedure, which a researcher may have followed while analyzing the research results of the research findings. Research methods may vary considerably from subject to subject and from science to science and each science generally follows a predefined and acceptable set of research methods to undertake different kinds of research activities. Some of the commonly followed research practices in the social sciences include.

- Sampling technique (this part shows how a study is replicable)
- Instruments used should be appended at the end
- Time frame for data collection
- Data analysis (SPSS, etc.)
- Statistical tests.

Review of Literature

Each kind of research activity undertaken in any given field of study traces its roots in the already available pool of research done previously by the researchers during different periods of time. Any researcher whosoever seeks to examine a particular aspect or phenomenon is not possible without identification of a research gap. This research gap can be identified only by going through, if not all, but a sizeable and latest related literature produced in any given field. By reviewing the already produced pool of literature, a researcher is able to formulate the basis of his own research in broader, clearer and acceptable way.

Review of literature gives an insight about the following areas:

- A comprehensive overview of prior research on a specific topic shows what is already known about the subject.
- Identify gaps-what is not known, creates background or provides rationale for new study or research.
- "The literature review is where a researcher identifies the theories and previous research which influence his /her choice of research topic and methodology." Ridley, 2012
- The importance of understanding, both as a researcher and as a reader of a research study, what is already known about a topic is primarily for doing high quality and useful research.
- "A substantive, thorough, sophisticated literature review is a precondition for doing substantive, thorough, sophisticated research." Boote and Beile, 2005

While reviewing the literature on any given topic, a researcher should take care of following few things.

- A researcher must be having complete control over the different research facets of his/her subject.
- The literature quoted in the review part of the study must uphold the undergoing study by presenting the facts and figures about the already available knowledge base in any given field and the gap which exists somewhere must warrant the need for undertaking that particular study
- Must know about the databases of your field.

Literature search strategy is equally an important component, which forms the basis of the literature review. It is always desirable that a researcher should be able to locate all the related literature published in his/her given field of research. In order to have a healthy and possibly exhaustive pool of literature available, there is far greater need that a researcher should design a comprehensive search strategy. A researcher must be aware of the following search techniques and the sources of information where from one can get the desired piece of information.

Search Techniques

- Search the database by keywords using filters to limit database, types of resources and time span.
- Simple search
- Advance search
- Library catalogue
- Interlibrary Loan
- Write down the search you conduct in each database so that you may duplicate them if you need.
- Use bibliographies and references of other research studies you find to locate more studies.
- Use citation management tools to keep track of your research citation.

Sources of Literature Search

Although, there are numerous source of information, but among both academic and scientific research community following are some popular sources of information more often consulted by both researchers and academicians.

- Academic journals (print + online)
- Books (print + online)
- Government reports
- Newspapers
- Classic–to the most up to date studies need to be included
- Peer reviewed content

Literature Search

Literature search should generally begin by reviewing the abstract of published research. If the seeker of information finds the abstract of a research paper informative or indicative to serve his/her review interest, then the researcher can decide about going through the whole paper. While going through a particular research article a researcher must never lose the sight to see the following questions.

- The research question
- The main idea, as what were the authors trying to find and convey
- Is the research funded by a source which could influence the findings?
- The methodology used to collect data

Review Writing

After completing the relevant information search and reviewing the related literature, then comes the time for review writing. Review writing is one of the important and integral components of research writing. The review should never be presented in wayward manner, an attempt should be made to reflect such relevant studies, which apart from being the latest in the field should also reflect the research developments sequentially. It is always important that the review part of the research should include:

- The most relevant and significant studies
- Be critical (Evaluate and discuss)
- Reflect the individual opinion of a researcher in the light of literature reviewed in correlation with the undergoing study
- The review should not be a mosaic work
- Discussion should be opened and closed with researcher own ideas
- Synthesize your own ideas
- Make sure to reflect that your argument was built upon the ideas of others
- Indicate which ideas were taken from whom
- Substantiate your view point.

Use Quotes Sparingly

Quite often, it has been observed that review part in most of the research articles reflects the opinion of other researchers with no individual observation of researchers about the earlier research findings. Literature should be put into quotes only if it warrants so, while as, a researcher should always aim to express the content and context, along with findings and opinions in his/her own words.

- Include research studies which support and oppose the aims and objectives of your work.
- Differentiate between fact and opinion.
- Conclusion of review of literature is bridged to the current study. It must be clear and concise about what was available in the literature, need not be very specific about the details. Gaps are identified in methodology or findings.

In Text Citations

It is a very common practice among the researcher across the world to cite different sources of information in their research study. Although, the researcher keeps citing the different source and the references to the same are generally given at the end of a research article or at the end of each page as a footnote. Many a time the need arises when it becomes imperative to cite along with reference page, etc. within the text, especially while putting text in verbatim. A reference also helps and guides a reader to consult those sources of information which may have been consulted by the researcher while undertaking any particular study.

Some of the commonly cited sources of information in the text include

- Direct quotes
- Paraphrases
- Words or terms unique to authors or research studies
- Historical, statistical or scientific facts
- Proverbs
- Any other common knowledge.

Results/Findings/Data Analysis

Results/findings/data analysis, etc. are some of the common names generally used for almost all sorts of empirical analysis. Data analysis part forms the main focal point of any research paper. It is this part or chapter around which the rest of research activity revolves. The analysis should always be undertaken in as simple way as possible so that same be easily understood the readers. There is no need to get into complex computations, which otherwise can be easily explained by performing simple and easy to understand expressions, duly supported by a well-structured theoretical analysis.

- This part of the research paper generally deals with the data analysis
- This section reports the research data-which has to be properly captured and organized
- Data can be presented through tables under various objectives.

Discussion

Discussion forms an important part of any research activity. The discussion in any research report should always be undertaken in the light of outcome one seeks from the data analysis. A researcher is supposed to ponder over various facets of research, he/she may have undertaken in the light of research findings. The researcher should evaluate his/her research findings by correlating them with the research findings already presented by other researchers in the similar area. A researcher is required to interpret each aspect of his/her research, in the light of the objectives of the study or the hypothesis framed.

Discussion of any researcher activity addresses each aspect be it, need, purpose or importance of the research activity, objectives, research methods or the limitations of research activity along with the scope of the study.

In brief, some of the key areas reflected in the discussion part of research article include.

- Interpretation has to be done in discussion part
- It starts with a link to the introduction of the study
- Settling down the research questions with the research findings.
- It refers to the literature review and demonstrates how the current study fits into the wider framework of current knowledge
- If it is supported or contradicted by other studies
- Does it add to the present base of knowledge
- How can it be used?
- Generalize the implications of the findings
- The focus should be beyond the original research questions.

Conclusion

Concluding part of a research article sums up research findings in view of the overall need, purpose and importance of a particular research study and addresses the very fundamental question of every research activity which a researcher raises in the title of research article itself. This is followed by closing the observation over the objectives and the hypothetical aspects of the study. Observation is also made about the scope for improvement of the study, which can further corroborate the present findings. Then ultimately the researcher puts facts straight, mostly based on the empirical facts on the trends which may have emerged from the data analysis. Also, the concluding part of a research article is more in contrast to the introductory part of research activity, it addresses mostly the queries which generally a researcher raises while introducing his/her research activity. The conclusion should be as brief as possible, but that should not compromise with the summing up an article in the best possible way. Research findings, social implications of research activity and other novel areas should be reflected along with the necessary recommendations, which may emerge during the course of the investigation.

Referencing

References form an important part of any research activity. Every research activity is generally based on an already available knowledge base in the given field and the literature which a researcher consults and quotes in his/her own work the sources are always supposed to be acknowledged in the form of references at the end of a research article. There is always a proper way of writing references and a researcher cannot take the liberty while acknowledging the work of other researchers. Even if we see, these days each journal has its own individual way of writing references. Every researcher is supposed to take care of the referencing style of each individual journal in which one desires to publish his/her research results.

Cover Letter

Each research article forwarded to an editor of a particular journal should always be supported with a cover letter this cover letter should clearly mention about the submission of the attached article for publication in that very particular journal. The cover letter should also clearly signify about the other aspects of the article like title, authors, institutional affiliation, designation, address, etc. It is equally desirable to mention in the cover letter that the article has neither been submitted to any other journal for publication nor is under consideration for publications elsewhere. Copyright and intellectual property rights are important areas of concern, which are taken care by the editors nowadays. A researcher is always required to submit an undertaking to the editor of a particular journal about the transfer of the copyrights and the claim of being the true author of the research article submitted for publication, upholding that there are no IPR and copyright issues involved the content submitted for publication.

BIBLIOGRAPHY

1. Boote DN, Beile P. Scholars before researchers: On the centrality of the dissertation literature review in research preparation. Educational researcher. 2005;34(6):3-15.

2. Supporting Research Writing: Roles and Challenges in Multilingual Settings. Using genre analysis and corpus linguistics to teach research article writing. 2013. pp. 55-71.

3. Calkins L. The art of teaching writing. Heinemann Educational Books Inc., 70 Court St., Portsmouth, NH 03801.

4. Crawley GM, O'Sullivan E. The Grant Writer's Handbook: How to Write a Research Proposal and Succeed 2016.

5. Creswell J. Research design: Qualitative, quantitative, and mixed methods approaches. SAGE Publications, Incorporated 2009.

6. Gosden H. Success in research article writing and revision: A social-constructionist perspective. English for specific purposes 1995;14(1):37-57.

7. Hayes JR, Flower LS. Writing research and the writer. American 1986;41(10):1106.

8. Isaac S, Michael WB. Handbook in Research and Evaluation, A Colection of Principles, Methods and Strategies Useful in Education and The Behavioral Sciences 1971.

9. Jalalian M. Writing an eye-catching and evocative abstract for a research article: A comprehensive and practical approach. Electronic Physician 2012;4(3):520-4.

10. Jones SR. (Re) Writing the Word: Methodological Strategies and Issues in Qualitative Research. Journal of College Student Development 2002;43(4):461-73.

11. Jordan RR. English for academic purposes: A guide and resource book for teachers. Cambridge University Press 1997.

12. Kerlinger FN, Lee HB. Foundations of Behavioral Research 1999.

13. Li Y. Apprentice scholarly writing in a community of practice: An intraview of an NNES graduate student writing a research article. TESOL Quarterly. 2007;41(1):55-79.

14. Modern Language Association of America. MLA handbook for writers of research papers. Modern Language Association of America 2010.

15. Mugenda OM, Mugenda AG. Research Methods: Quantitative & Qualitative Approaches. African Centre of Technology Studies. 1999.

16. Myers MD. Qualitative research in information systems. Management Information Systems Quarterly, 1997;21(2):241-2.

17. Perneger TV, Hudelson PM. Writing a research article: advice to beginners. International Journal for Quality in Health Care. 2004;16(3):191-2.

18. Pratt MG. From the editors: For the lack of a boilerplate: Tips on writing up (and reviewing) qualitative research. Academy of Management Journal. 2009;52(5):856-862.

19. Ridley D. The literature review: A step-by-step guide for students. Sage 2012.

20. Šesták Z. Holliday, A.: Doing and Writing Qualitative Research. Photosynthetica. 2008;46(2):261-2.

21. Swales JM, Feak CB. Academic writing for graduate students: Essential tasks and skills (Vol. 1). Ann Arbor, MI: University of Michigan Press 2004.

22. Taylor SJ, Bogdan R, DeVault M. Introduction to qualitative research methods: A guidebook and resource. John Wiley & Sons. Chicago 2015.

23. Teale WH, Sulzby E. Emergent Literacy: Writing and Reading. Writing Research: Multidisciplinary Inquiries into the Nature of Writing Series. Ablex Publishing Corporation, 355 Chestnut St., Norwood, NJ 07648 1986.

24. Trochim WM, Donnelly JP. Research methods knowledge base 2001.

25. Turabian KL. A manual for writers of research papers, theses, and dissertations: Chicago style for students and researchers. University of Chicago Press 2013.

26. University of Birmingham. Department of English Language and Literature, Sinclair, J. M. (1972). The English used by teachers and pupils: final report to SSRC for the period 1st September 1970 to 31st August 1972. University of Birmingham, Department of English Language and Literature.

27. Voss C, Tsikriktsis N, Frohlich M. Case research in operations management. International Journal of Operations & Production Management. 2002;22(2):195-219.

28. Webster J, Watson RT. Analyzing the past to prepare for the future: Writing a literature review. MIS Quarterly. 2002. xiii-xxiii.

Chapter 5

Cite While You Write: A Practical Approach with EndNote

INTRODUCTION

It has been observed that researchers face a range of problems in acknowledging the research contribution of other researchers in their studies. Referencing has been subject of discussion for various reasons and the foremost being that citations are the only method as on date available with us, which is being used as a parameter to judge the quality of a research article or any other research publication. Therefore, there is always far greater need that one should adhere to a referencing style, which is advocated and acceptable by a particular journal or a publisher. The underlying discussion revolves round the technological offing, wherein this tiring work has been made easier with the help of reference management software. There are different reference management software readily available in the market. Some of than are proprietary and some are non-proprietary, which we commonly refer as open source software.

This chapter deals with the reference management software EndNote, which is a proprietary software and people can purchase it to avail its services; however, the trial version of the software can be accessed for limited use. EndNote is a very popular software among researcher to "cite while write", has more or less become a buzz word nowadays. The software is generally used while writing a research article, thesis, dissertation or any other type of research work, which needs to be corroborated by proper referencing. The software can be easily used to organize literature, to cite sources and in compiling a bibliography, etc. without wasting time. Besides, there are numerous other features which can be easily explored to make good use of the software. The present discourse lasts around the installation and use of EndNote, the reference management software, with some step by step screen shots to learn and understand the basic nitty-gritty of the software and its use.

References constitute the important and an integral part of any research work, be it research paper or report writing. The purpose of acknowledging the others work is to give due credit to the authors, who may have contributed significantly in any given field of study and forms the important correlation with your current study. There are various methods to cite others work in one's own work, what we often refer as citation styles or reference formats. However, if a person quotes others work by whatever means, but fails to cite it or did not acknowledge the others work, the quoted work will be considered as plagiarism. There are many reference management tools available for capturing the right source, citing and making a list of references.

The researchers and the academicians are always under the constant pressure to write and cite, which is a time-consuming exercise, thereof the follow-up work like citing different sources is always being dubbed as hectic and boring aspect of the work. Collection and management of exact citations of sources of information is more time-consuming and frustrating task for a researcher, as a good portion of their writing is wasted for the want of proper reference writing style, which is more often managed manually. Sometimes importing references from online and offline databases without journal-specific format are later required to be formatted as per journal requirement. Any inefficiency or inability on the part of the researcher or author to not to cite the others work properly will amount to discrediting the others work, goes against intellectual effort of original authors and amounts to plagiarism.

Any work related to literature, history, current events and many other fields, direct quotes may be essential for the full discussion of a subject. In science, normally we do not find any direct quote, as most of the research undertaken in the pure sciences is generally based on lab work, hence may not warrant discussion in the light of earlier works. Similarly, in student papers there is no reason at all to include direct quotes, except in the case when the student does not understand the concept and uses the quote to avoid having to explain it. As a rule, do not use direct quotes in a scholarly technical paper. Your own thoughts must be expressed, not those of someone else. (Caprette DR)

In this chapter the discussion lasts around capturing the right references, creating a database and citing in proposed publication and later on the format as per journal-specific for publication.

KEY ISSUE OF A RESEARCH PUBLICATION

The following major component should present in a research/review publication. (Middelberg A)

- Title
- Abstract
- Introduction
- Methodology
- Result
- Discussion/Conclusion
- References

All the major components the references should be verified in the following ways

- If the article builds upon previous research, does it include the reference of the consulted work appropriately?
- Are there any important works that have been omitted?
- Are the references accurate?
- Are the references properly styled?
- Are the references in proper format?

Here the issue is to manage the quality and quantity of references available for researcher and how to incorporate in a research article.

WHAT IS CITATION?

A "citation" is the way you tell your readers that certain material in your work is borrowed from other sources. It also gives your readers the information necessary to find that source again. (What is citation) Citation includes the following:

- Information about the author
- The title of the work

- The name and location of the company that published your copy of the source
- The date your copy was published
- The page numbers of the material you are borrowing.

Citations symbolize the association of scientific ideas. The references that research authors cite in their papers, make explicit links between their current research and prior work in the scientific literature archive. (Clarivate Analytics).

In short, we can say the acknowledgement of a publication receives from another publication is called citation.

WHAT IS A REFERENCE?

Reference is acknowledging the sources of information, view and ideas that are used in your publication. Reference is the confirmation that the documents have been consulted and due credit has been given to the author.

Works cited is also referred to as references. The terms are generally used interchangeably, but the fundamental difference between the two is, any acknowledgment of others works mentioned at the end of a research article is mentioned as reference. The same reference when seen from the perspective of the original author or any other peer researcher is viewed as citation. Hence, your references are others citations. Normally, references are listed either in the alphabetical order or as per their reference order in the document. Cited reference and bibliography is not the same. In the works cited, you only list items you have actually cited. In a Bibliography you list all of the material you have consulted in preparing your essay whether or not you have actually cited the work. (Gibaldi J)

People generally use the terms bibliography and references interchangeably, but the fundamental difference between the two is that, bibliography lists all the materials that have been consulted while writing an essay or a book. References, on the other hand, are those that have been referenced in your article or book. (Answers Corporation)

WHY REFERENCE?

Verify the original work and enable the readers to know the fact about your written work and most importantly to avoid plagiarism are the main reason for referencing.

References must be provided whenever you use someone else's opinions, theories, data or organization of material. You need to reference information from books, articles, videos, computers, other print or electronic sources and personal communications. A reference is required, if you: (Division of Teaching & Learning Services).

- Quote (use someone else's exact words)
- Copy (use figures, tables or structure)
- Paraphrase (convert someone else's ideas into your own words)
- Summaries (use a brief account of someone else's ideas).

Harris R. has developed a simple flow chart to assist researcher to determine which research reference should be cited and which do not cite during the writing.

Flowchart 1. Robert Harris's Flowchart for Citation

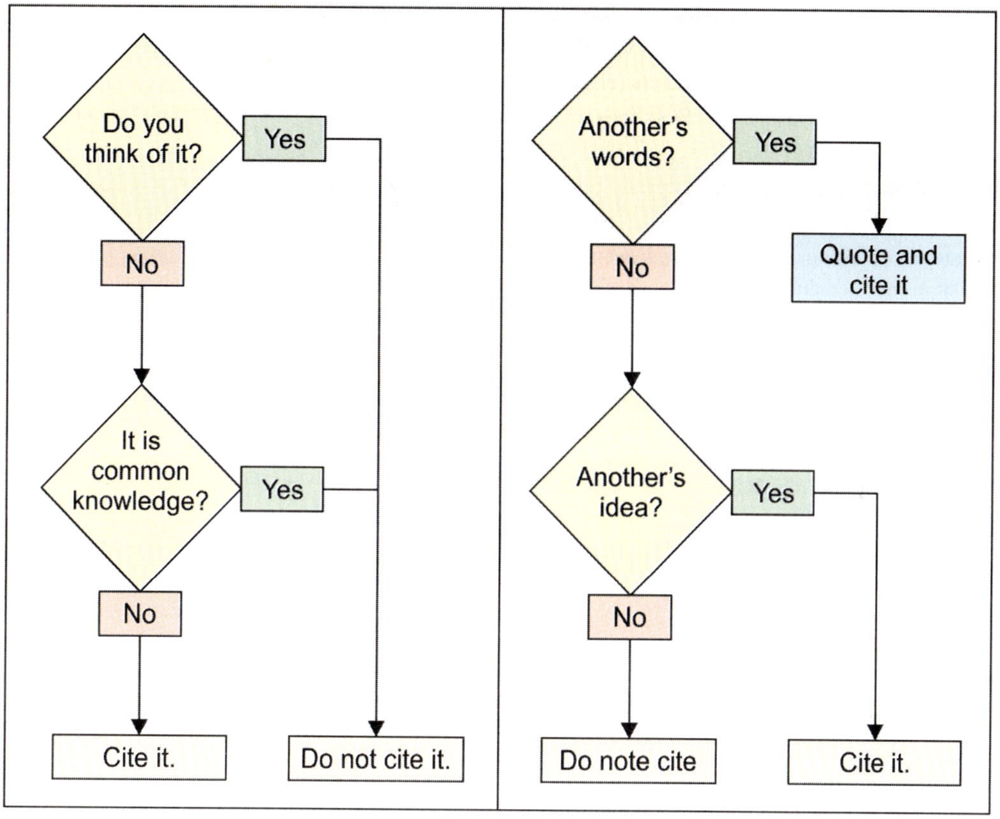

```
Do you think of it?   Yes
    │
    No
    │
It is common knowledge?   Yes
    │
    No
    │
Cite it.          Do not cite it.

Another's words?   Yes
    │                │
    No          Quote and cite it
    │
Another's idea?   Yes
    │               │
    No              │
    │               │
Do note cite      Cite it.
```

Source: Harris: *The Plagiarism Handbook*, 2001; pp155, 158

REASONS FOR GIVING CITATIONS

E Garfield said the issue of "when to cite" is closely related to questions about the "way to cite" and has given fifteen major reasons for citation. (Garfield E.)

1. Paying homage to pioneers
2. Giving credit for related work (homage to peers)
3. Identifying methodology, equipment, etc.
4. Providing background reading
5. Correcting one's own work
6. Correcting the work of others
7. Criticizing the previous work
8. Substantiating claims
9. Alerting researchers to forthcoming work
10. Providing leads to poorly disseminate, poorly indexed or uncited work

11. Authenticating data and classes of fact, physical constants, etc.

12. Identifying original publications in which idea or concept was discussed

13. Identifying the original publication describing and eponymous concept or term

14. Disclaiming work or ideas of others (negative claims)

15. Disputing priority claims of others (negative homage).

REFERENCE STYLE

There are many citing and reference listing styles. Author must follow the instructions as per journal to which an author is submitting a paper. References must be in the style required by the journal. More than 4,500 bibliographic styles are listed in the current version of EndNote. These styles are specific as per journal requirement. Some journals and even organizations are using some popular styles, likes APA (American Psychological Association), Chicago Style (The University of Chicago Press), MLA Style (Modern Language Association) and Vancouver system.

REFERENCE MANAGEMENT SOFTWARE

Reference management software are sometime referred or call as citation management or bibliographic management software. These software are designed to capture and manage the references and easy to use in a document for proper referencing. These software also help to create various bibliographies, which prove quite useful in undertaking bibliometric studies. These reference software prove useful in capturing references online as well as offline. The references are automatically created at the end of the publication the moment author cites a particular document within the text with the help of reference management software.

These software packages are integrated with most of the usually used word processing application software generally used for writing the research papers. Earlier version of software packages used to need sound command over these software to run the task, but now most of format citation and reference are fixed automatically with minimal effort. These software packages have their own database and are able to generate various bibliographic reports and list in the different formats, styles as per standards set by the publishers and other journal styles. The software also provides the facilities to create a new style, format and also edit the pre-existing style. The major feature is the facility to import and export references in each other's format along with the automatic web-based import features.

Reference management software does not do the same job as a bibliographic database, which tries to list all articles published in a particular discipline or group of disciplines; examples are those provided by Ovid Technologies (e.g. Medline), the Institute for Scientific Information (e.g. Web of Knowledge) or mono-disciplinary learned societies, e.g. the American Psychological Association (PsycINFO). These databases are large and have to be housed on major server installations. Reference management software collects a much smaller data, of the publications that have been used or are likely to be used by a particular author or group, and such a database can easily be housed on an individual's personal computer. (Wikipedia 2010).

There are many software available for the management of references, some of them are proprietary software and many other non-proprietary, available in GPL. In early 1980s the software were purely proprietary, but later on post 2001 many software were made available in Open Source Software (OSS). A comparative chart of some of the proprietary and OSS reference management software is given in Table 1. (Wikipedia 2015)

Table 1: Comparative study of reference management software

Software	Developer	First public release	Cost (USD)	Free software	License	Notes
BibDesk	BibDesk developers	2002-04	Free	Yes	BSD	BibTeX front-end + repository
EndNote	Clarivate Analytics	1988	US$299.95[1]	No	Proprietary	The web version EndNote Basic (formerly, EndNote Web) is free of charge
JabRef	JabRef developers	2003-11-29	Free	Yes	GPL	Java BibTeX manager
KBibTeX	KBibTeX developers	2005-08	Free	Yes	GPL	BibTeX front-end
Mendeley	Elsevier	2008-08	Free / Online storage free up to 2 GB	No	Proprietary (OS API clients exist)	Desktop and Web components, Windows, Linux, OS X, iPhone and iPad
Papers	Springer	2007	US$79[1]	No	Proprietary	Search repositories from interface; supports plug-ins, Universal iOS app
ReadCube	Labtiva	2011-10	Free desktop and mobile app / Cloud Storage $5/month, $50/year	No	Proprietary	Desktop (Mac/PC) and Web components, integrated web search, enhanced PDF reader, iOS App, Android
Reference Manager	Thomson Reuters	1984	US$239.95[1]	No	Proprietary	Network version available; built-in web publishing tool
Referencer	Referencer developers	?	Free	Yes	GPL	BibTeX front-end
RefWorks	RefWorks / ProQuest	2001	US$100 per year	No	Proprietary	Web-based, browser-accessed, centrally hosted program
SciRef	Scientific Programs	2012	US$38.90 / Free trial version	No	Proprietary	
Zotero	Roy Rosenzweig Center for History and New Media at GMU	2006-10-05	Free / Online storage free up to 300 MB	Yes	AGPL	Firefox extension or stand-alone with connectors for Firefox, Chrome and Safari. Web-based access to reference library also available.

ABOUT THE ENDNOTE

The EndNote reference management software was developed by the Institute of Scientific Information (ISI) in 1988. Now it is with Thomson Reuters. The Latest version of EndNote is X4. The software can store and organizes citations found from many sources and one can insert these citations into a word document, which automatically formats references according to a predefined citation style.

EndNote is a reference management software application used by researchers, academicians and librarians to organize citations, references, bibliography and many other forms of scholarly work. EndNote can be linked with word-processing applications such as Microsoft Word to further automate the citation processes according to a predefined citation style. With this software linkage, users can easily extract or update references between Word and EndNote without changing the citation database itself.

FLOW PROCESS (FLOW CHART 2)

The flow digram demonstrates the thing to manage the references systematically. The flow digram of the EndNote consisted four phases, Reference entry manually or search from the web and then save; Import these save references into EndNote software; Export to Microsoft word during the writing; and finally create and reformat the desire citation and reference over 5000 styles.

Flowchart 2. Flowchart of EndNote

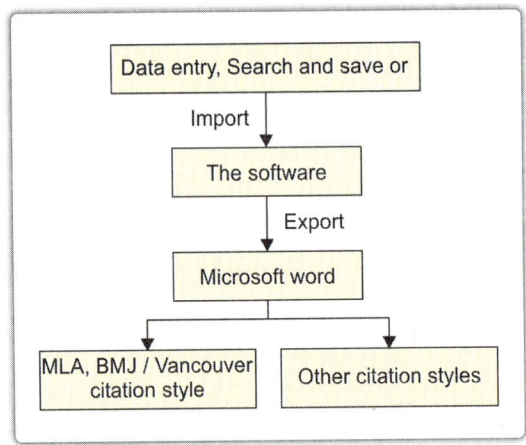

SYSTEM REQUIREMENT

Following software and hardware are required for installation and operation of the EndNote software.

Software Requirements

- Windows XP with SP3 or Vista
- Microsoft Word 2003, 2007 or 2010

Hardware Requirements

- Pentium 450 MHz or faster
- At least 180 MB hard disk space
- Minimum of 256 MB RAM
- Internet or LAN connection.

HOW TO GET ENDNOTE

The software is available in different access formats, be it as a single user, network version and also web based. The cost of software is given on their website (www.endnote.com). The trial version of the software is also available to download for a download and can easily used for 30 days. Upon finding the software interesting one can also download its paid version having the file size around 85.8 MBs.

HOW TO INSTALL?

Software can be obtained through purchase or can be downloaded directly from the internet (Demo Version for 30 Days). After getting the software, first it should be installed on your PC or Laptop **(Fig. 1).**

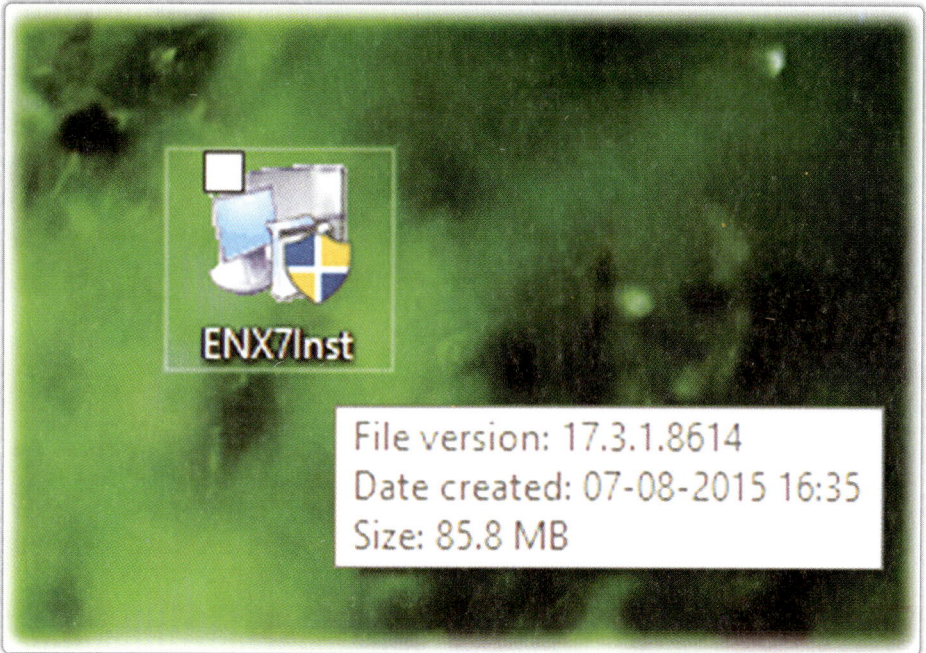

Fig. 1: EndNote Installer File (Windows)

To install the EndNote on your Windows Desktop need to run the downloaded or purchased installer file and follow the installation instructions.

Click on "Next > (button)

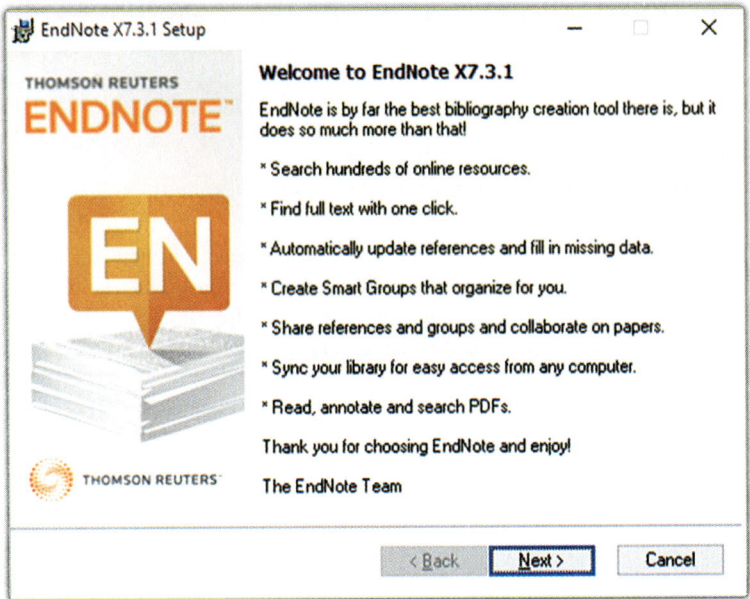

Fig 2. EndNote installation process

Source: EndNote X7.3.1 (Thomson Reuters) Recent versions are now with Clarivate Analytics

Click on "Next > (button)

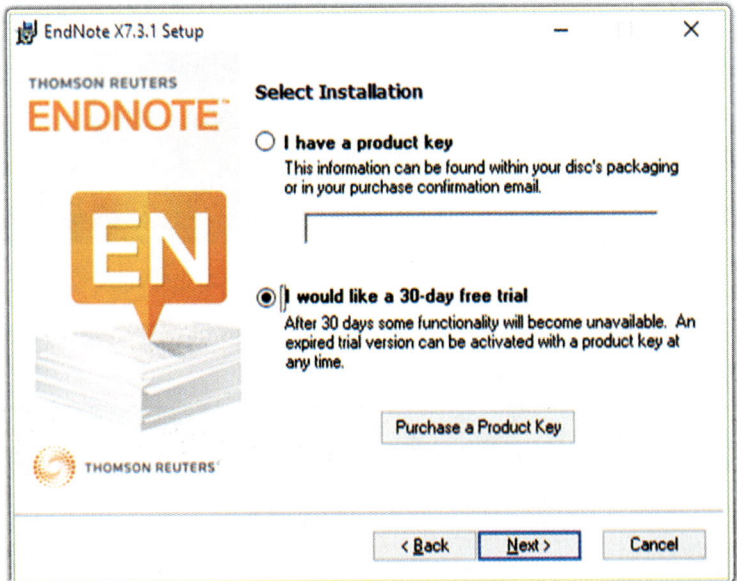

Fig 3. EndNote installation process

Source: EndNote X7.3.1 (Thomson Reuters) Recent versions are now with Clarivate Analytics

Click on "Next > (button)

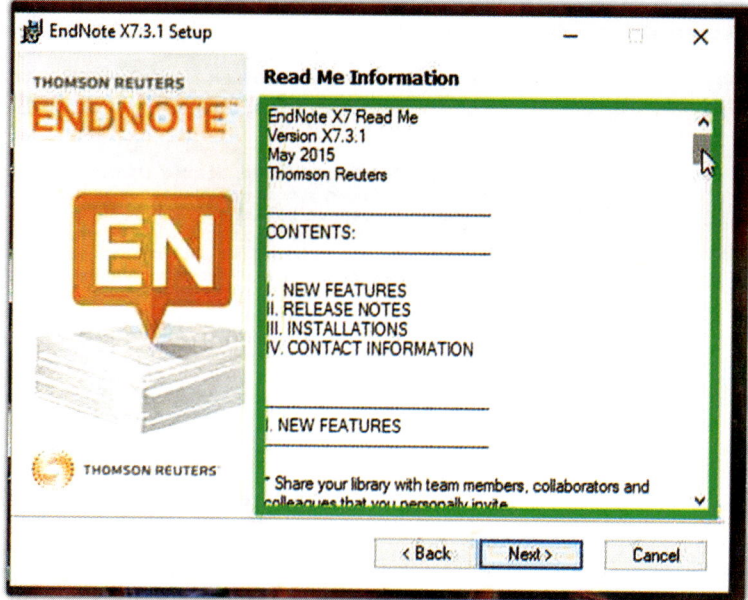

Fig 4. EndNote installation process

Source: EndNote X7.3.1 (Thomson Reuters) Recent versions are now with Clarivate Analytics

Click on "Next > (button)

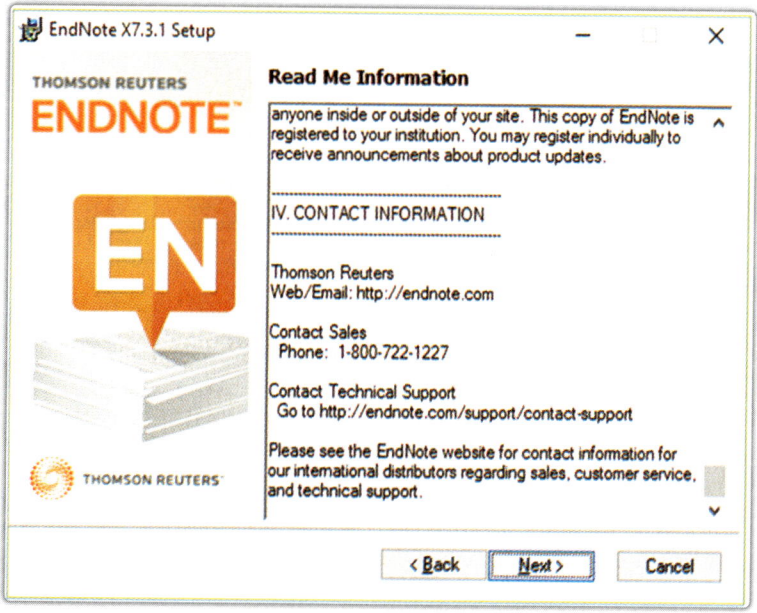

Fig 5. EndNote installation process

Source: EndNote X7.3.1 (Thomson Reuters) Recent versions are now with Clarivate Analytics

Accept the license agreement and Click on "Next" (button)

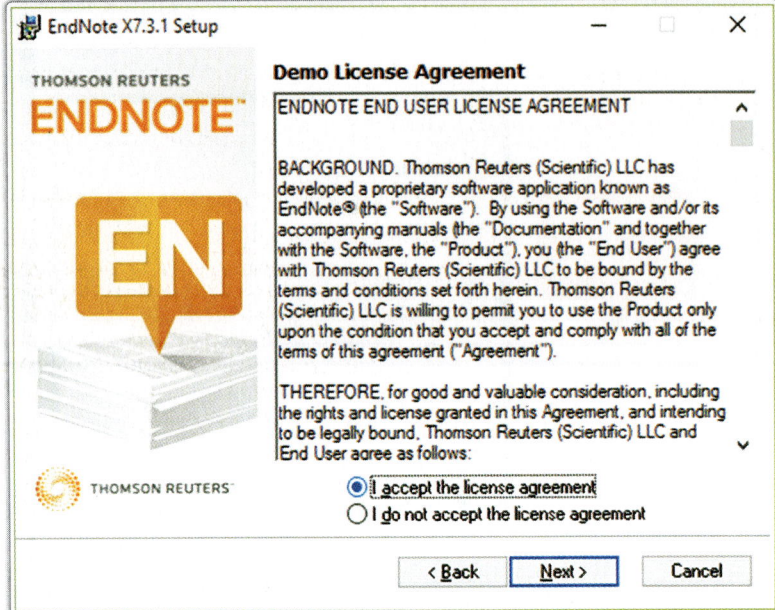

Fig 6. License agreement

Source: EndNote X7.3.1 (Thomson Reuters) Recent versions are now with Clarivate Analytics

Click on "Next > (button)

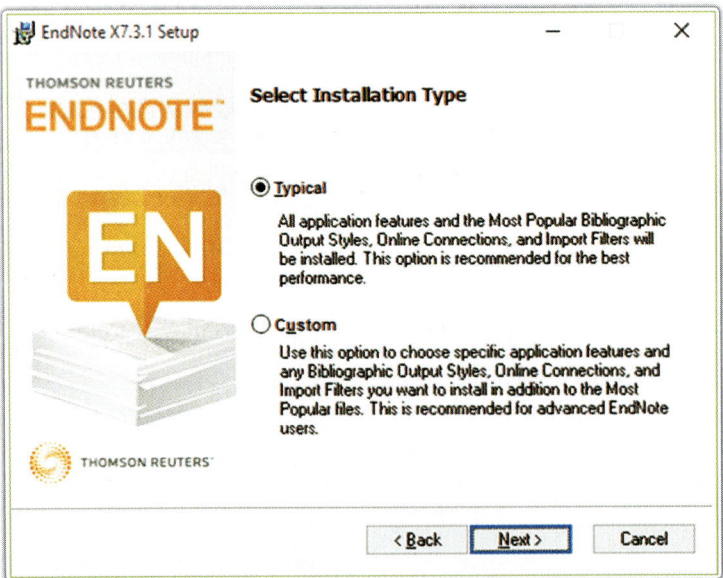

Fig 7. EndNote installation process

Source: EndNote X7.3.1 (Thomson Reuters) Recent versions are now with Clarivate Analytics

Select the default folder and click "Next." Click on "Next > (button)

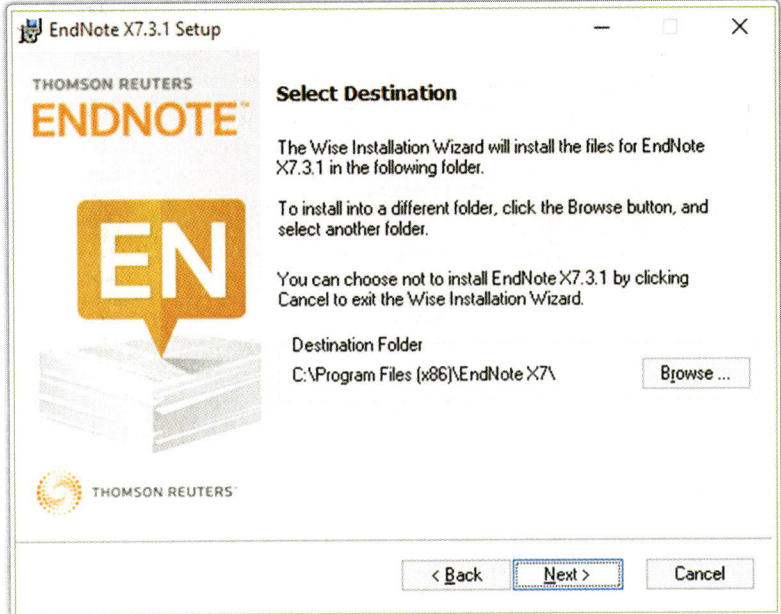

Fig 8. EndNote installation process

Source: EndNote X7.3.1 (Thomson Reuters) Recent versions are now with Clarivate Analytics

Click on "Next > (button)

Fig 9. EndNote installation process

Source: EndNote X7.3.1 (Thomson Reuters) Recent versions are now with Clarivate Analytics

Click "Finish" to finalize the installation after the updating the process of "Updating System"

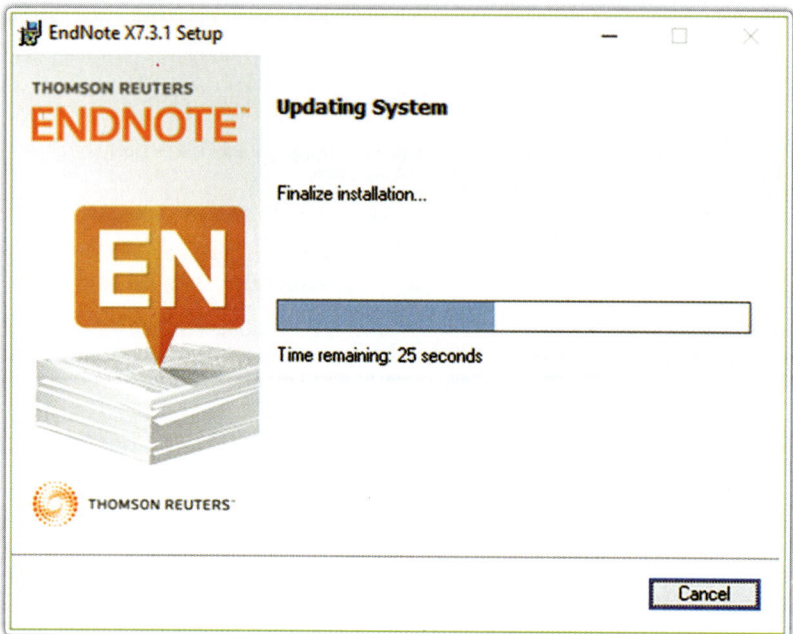

Fig 10. EndNote installation process

Source: EndNote X7.3.1 (Thomson Reuters) Recent versions are now with Clarivate Analytics

Fig 11. EndNote installation process

Source: EndNote X7.3.1 (Thomson Reuters) Recent versions are now with Clarivate Analytics

You should see "EndNote" under the 'All Programs' after clicking on 'Start'.

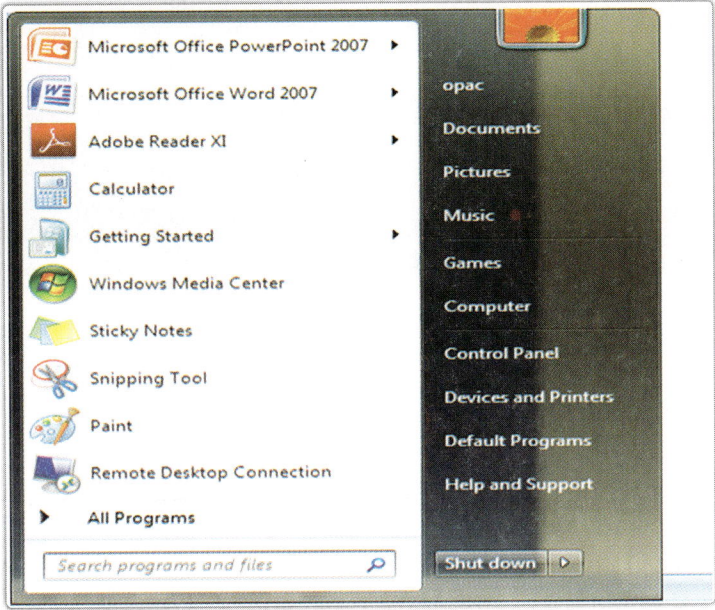

Fig 12. All programs in MS Windows

After selecting 'All Programs', you should see 'EndNote' in the program list.

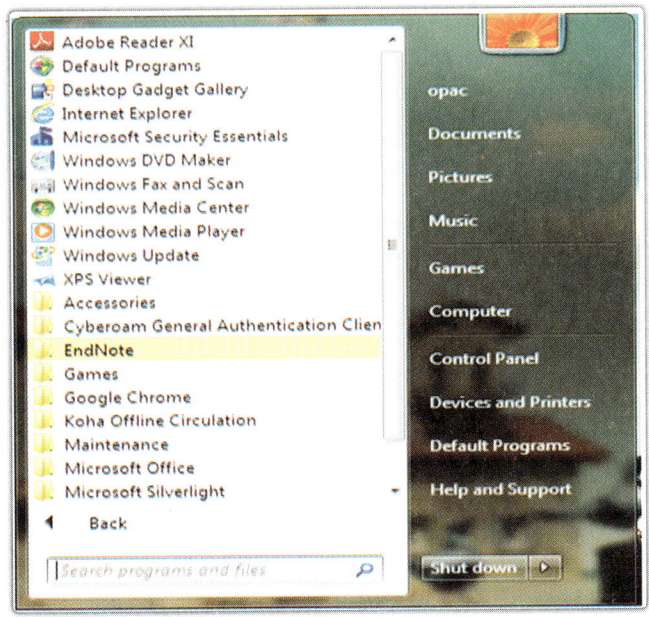

Fig 13. Endnote folder under all programs

Click on the "Endnote" in the list under the folder "Endnote" to open the program.

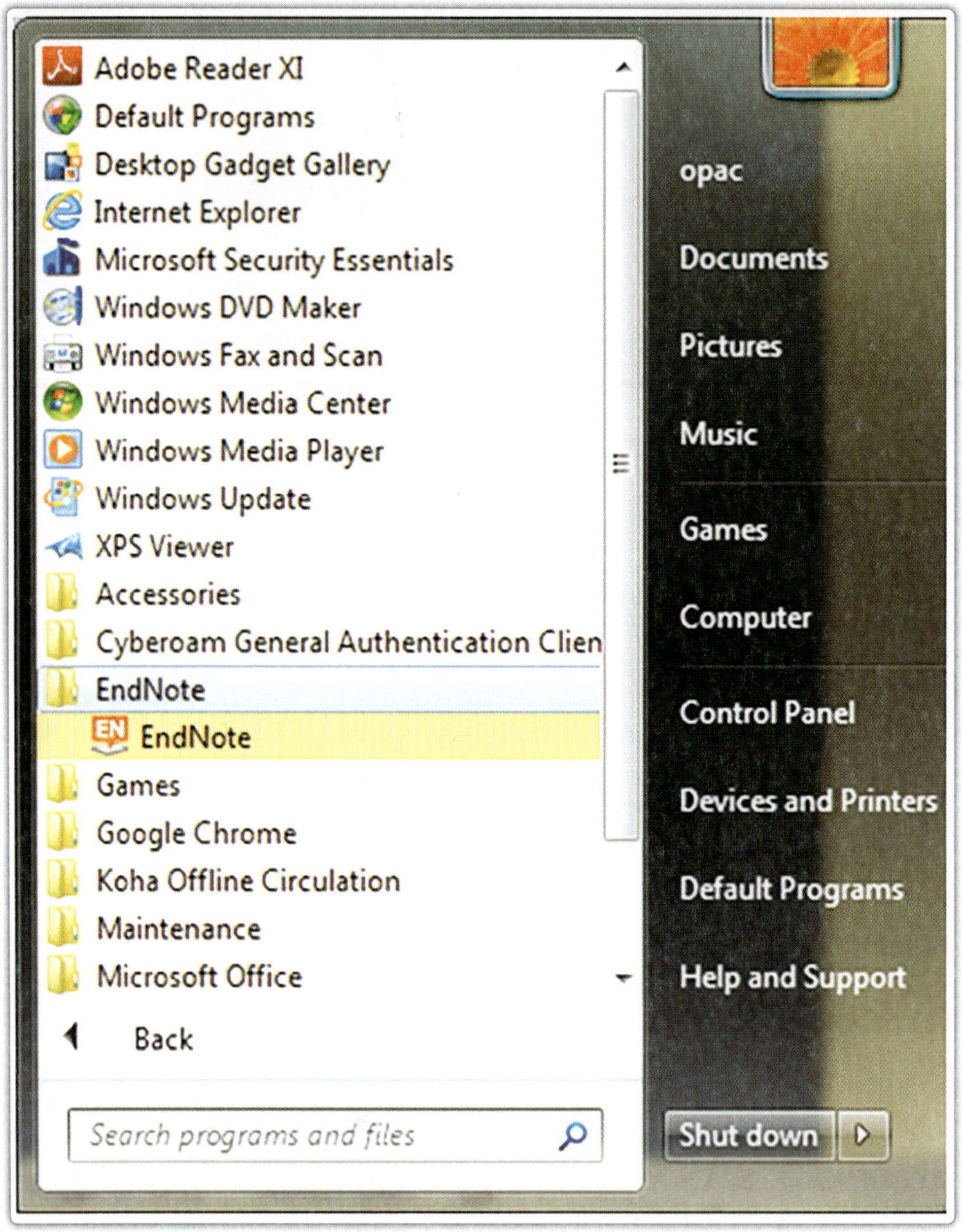

Adobe Reader XI

Default Programs

Desktop Gadget Gallery

Internet Explorer

Microsoft Security Essentials

Windows DVD Maker

Windows Fax and Scan

Windows Media Center

Windows Media Player

Windows Update

XPS Viewer

Accessories

Cyberoam General Authentication Clien

EndNote

    EN EndNote

Games

Google Chrome

Koha Offline Circulation

Maintenance

Microsoft Office

◀   Back

Search programs and files   🔍

opac

Documents

Pictures

Music

Games

Computer

Control Panel

Devices and Printers

Default Programs

Help and Support

Shut down ▷

Fig 14. Endnote program with in endnote folder

When the open "EndNote" program in the first time, mark the bubble "I accept the license agreement" to acknowledge that you have read and accept all terms of the license agreement. Then click "Next".

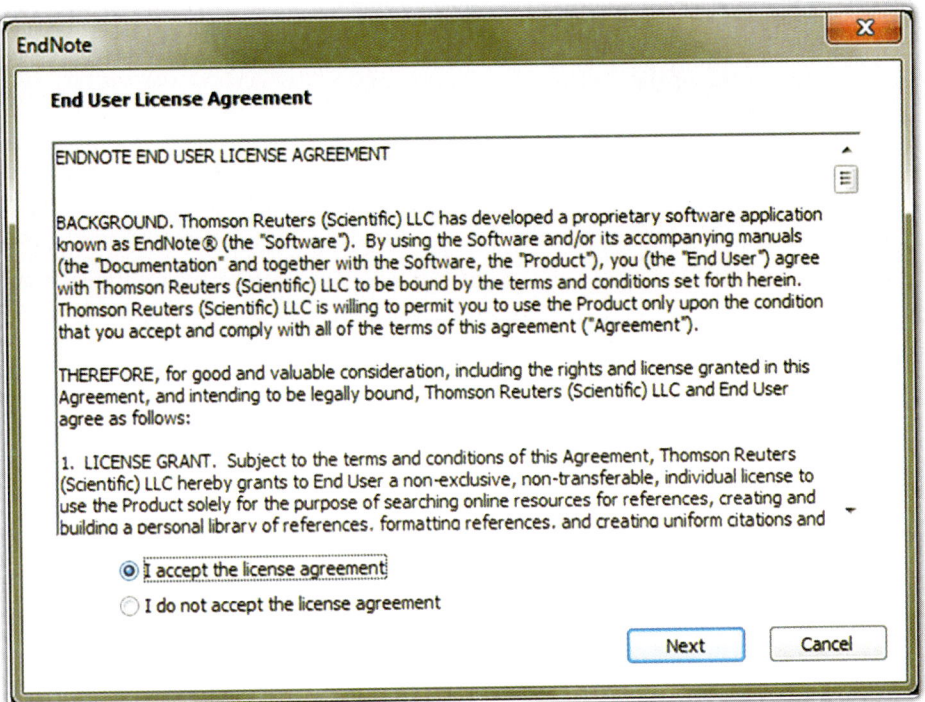

Fig 15. EndNote user license agreement

Select "No" to use the trial version

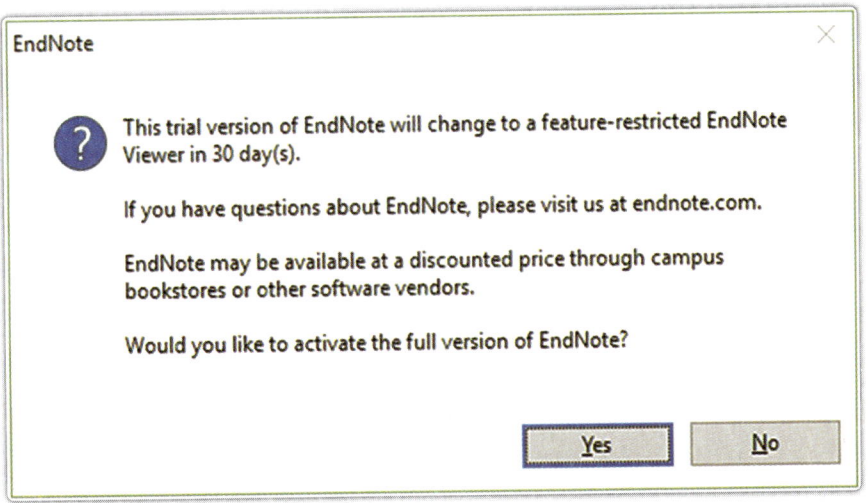

Fig 16. EndNote - Selection of trail vs full version

Fig 17. EndNote - Desktop screen

Source: EndNote X7.3.1 (Thomson Reuters) Recent versions are now with Clarivate Analytics

CREATE AN ENDNOTE LIBRARY / CREATING A REFERENCE FILE

The Sample library is placed in the EndNote Examples folder during installation and is used in the exercises in the EndNote Getting Started Guide. For creating a new file just click on File then new.

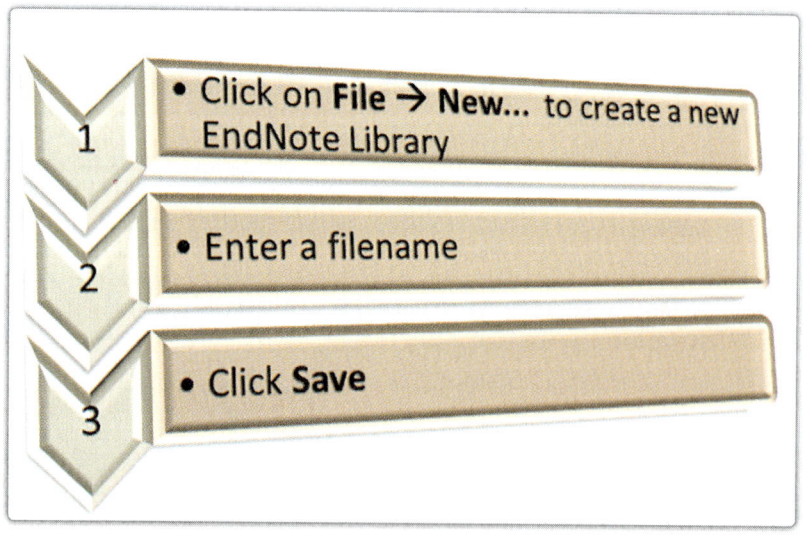

Fig 18. EndNote - Method to create first EndNote library

Source: EndNote X7.3.1 (Thomson Reuters) Recent versions are now with Clarivate Analytics

Click on File- New. To create a new EndNote Library

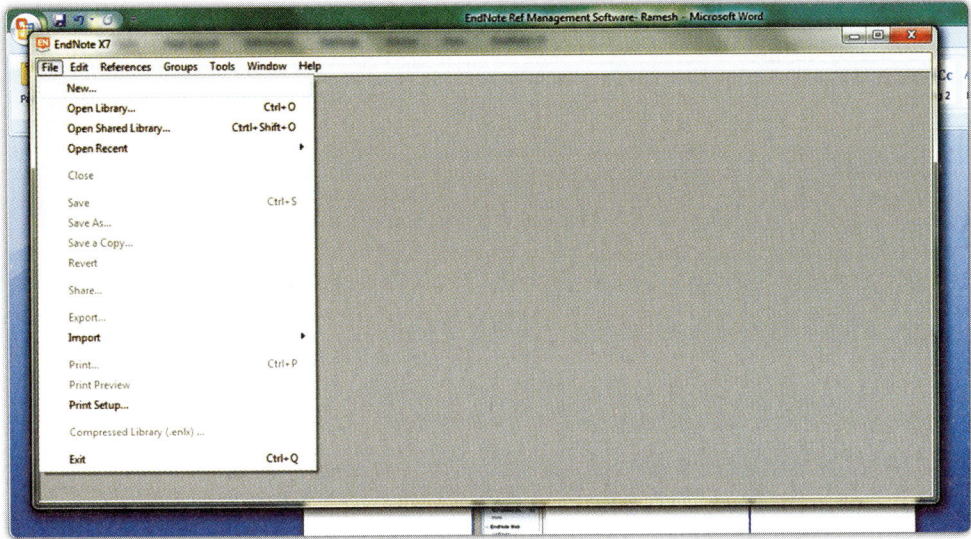

Fig 19. EndNote -File menu

Source: EndNote X7.3.1 (Thomson Reuters) Recent versions are now with Clarivate Analytics

Enter a File Name and Click Save

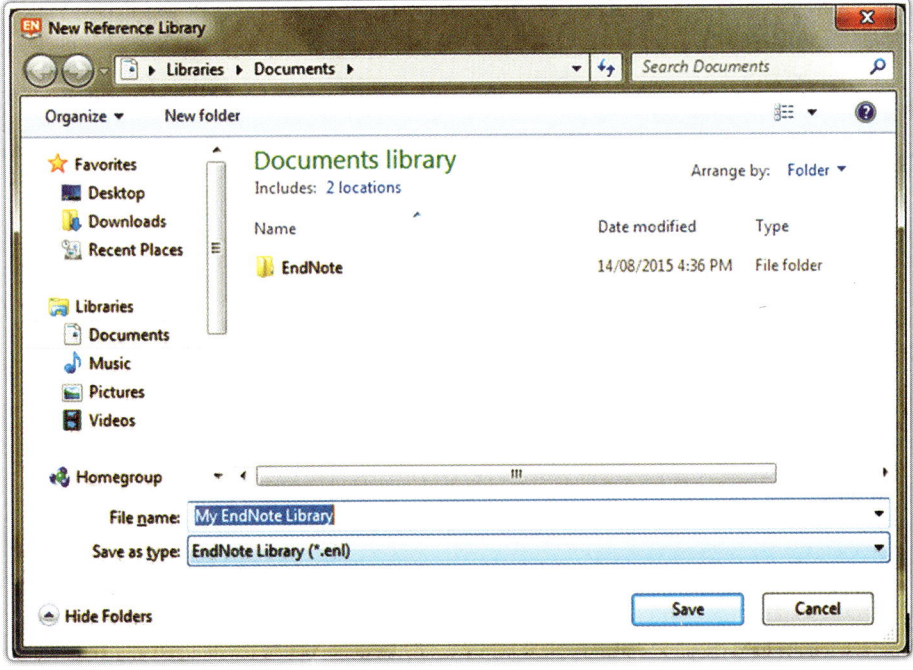

Fig 20. Endnote -Save Windows

After follow the basic steps of the default installation wizard and complete the installation. The following first screen will be displayed.

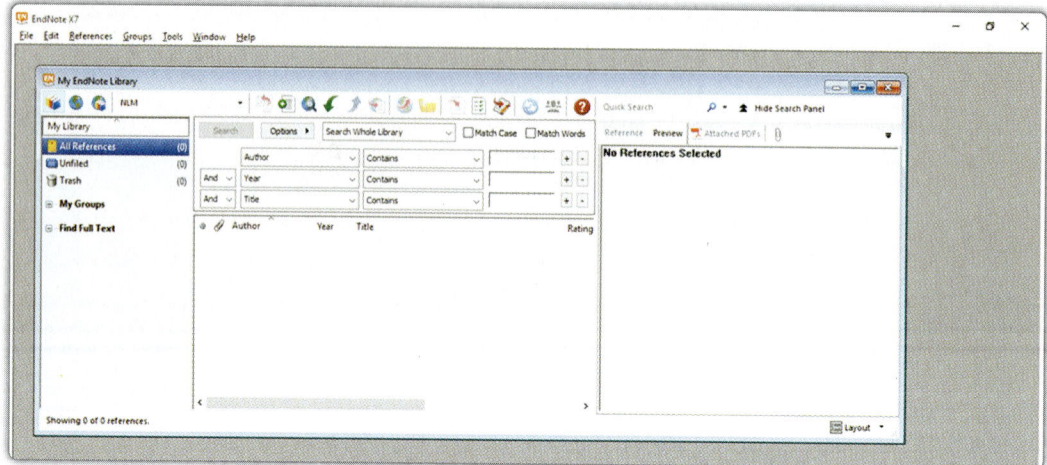

Fig 21. EndNote - My library

Source: EndNote X7.3.1 (Thomson Reuters) Recent versions are now with Clarivate Analytics

It allows for creation of an unlimited number of personal databases, which you can call libraries in EndNote. You can allow an unlimited number of libraries to be opened at one time. Each library can contain an unlimited number of references and each record can contain up to 52 fields (e.g., Author, Title, etc.) The Thomson Reuters EndNote Styles collection contains more than 4500 bibliographic styles for a variety of subject disciplines (Thomson Reuters (ISI) EndNote Class Outline).

There are standard formats for filling out a simple template that has the proper fields displayed for a given type of reference. There are 46 customizable reference types, covering a variety of materials from Ancient Texts to Web Pages, each with up to 52 fields for entering references. By using EndNote's one can simply open more than 3,900 predefined connection files online. One can access hundreds of remote bibliographic databases, including Web of Science, Ovid, PubMed, the Library of Congress, and university card catalogues from EndNote. Connect to data sources worldwide—EndNote provides MARC formats that support native language libraries around the world. Search remote bibliographic databases using EndNote's simple search window—which is very interactive and useful for locating specific references. Export references directly from Web of Science, Highwire Press, Ovid, OCLC, ProQuest and more. Save and load search strategies at the click of a button. Drag and drop references between EndNote libraries in one simple step. No additional importing required (Thomson Reuters. EndNote Information).

COLLECT REFERENCES/CITATIONS

- Method 1: Search in databases and export citations through any online/offline database (e.g. PubMed/Medline, etc.)
- Method 2: Search Library of Congress/PubMed/Web of Science/ProQuest/EBSCO/Gale/OvidSP and many more directly in EndNote
- Method 3 : Manually enter a Reference

Method 1: Search in databases and export citations through any online/offline database (e.g. PubMed/Medline etc.)

Perform search in PubMed

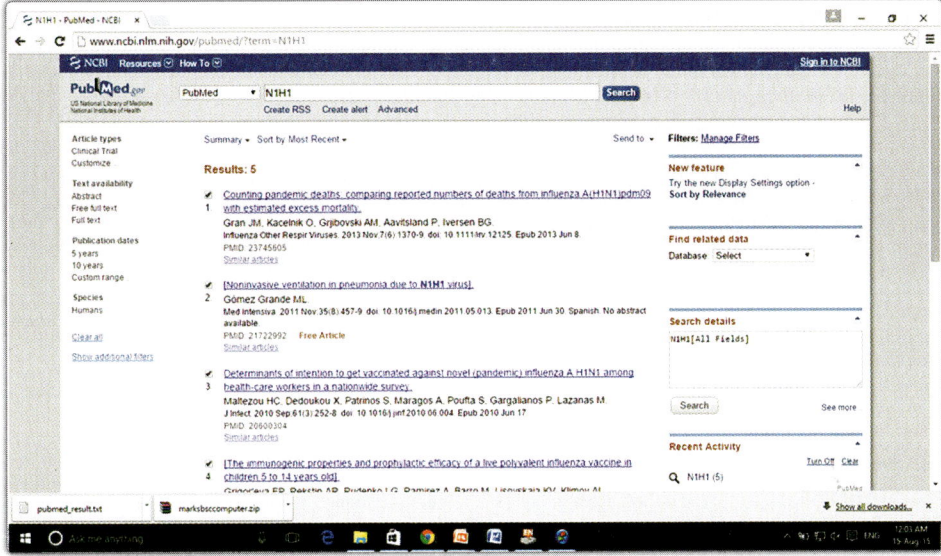

Fig 22. PubMed search windows

After Complete your query on PubMed (i) mark the references you would like to add to the file to export, (ii) then click on the "Send to" link to send the selected references, and mark "File" under the list of "Choose Destination", (iii) select MEDLINE as the Format, sorting preference and (iv) click the "Create File" button.

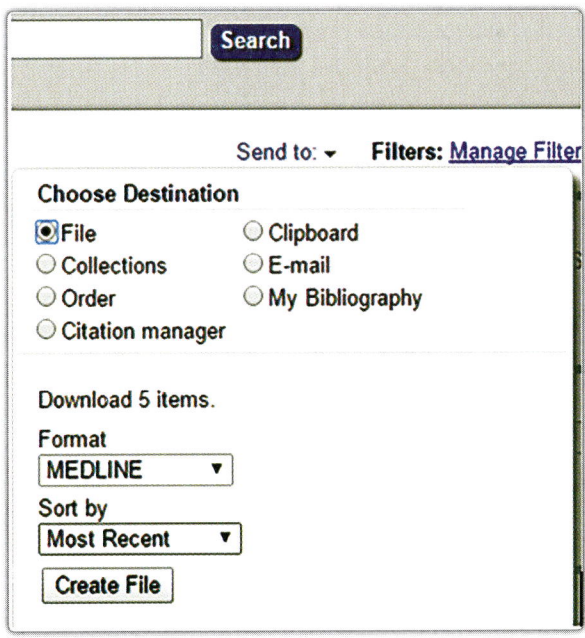

Fig 23. PubMed references - Export

v.   Save the file as a .txt file

Import into EndNote

- Open an EndNote library previously created.
- To import references, click on Import button.
- A dialogue box "Import" appears:

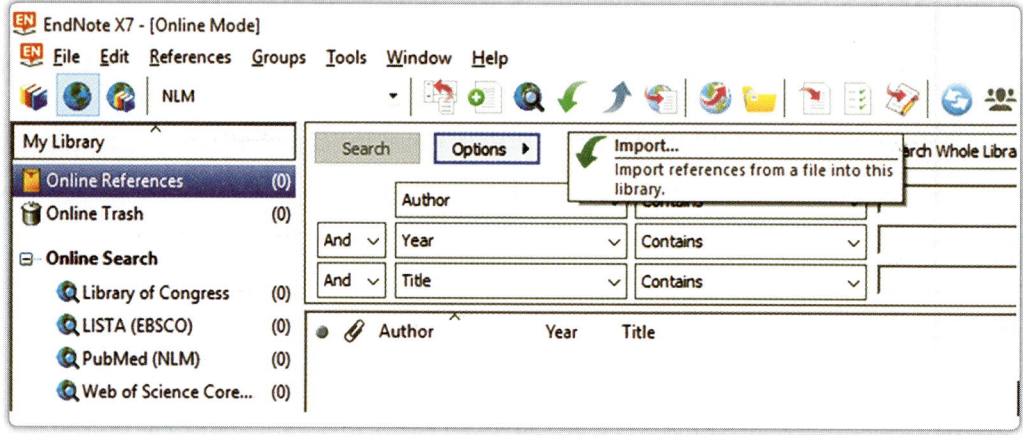

Fig 24. EndNote - Import menu

Source: EndNote X7.3.1 (Thomson Reuters) Recent versions are now with Clarivate Analytics

i.   **Import data file:** Browse for the saved file from PubMed

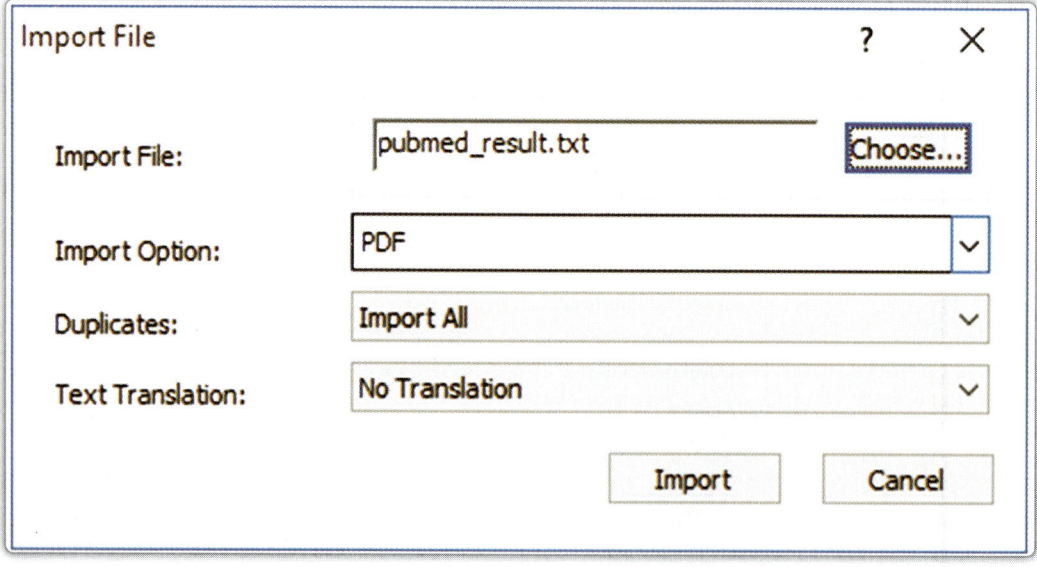

Fig 25. EndNote - Pop-up windows under the import menu

Source: EndNote X7.3.1 (Thomson Reuters) Recent versions are now with Clarivate Analytics

ii. **Import option:** Click **other filters.** Look for the database name **PubMed (NLM)**. Click **Choose.**

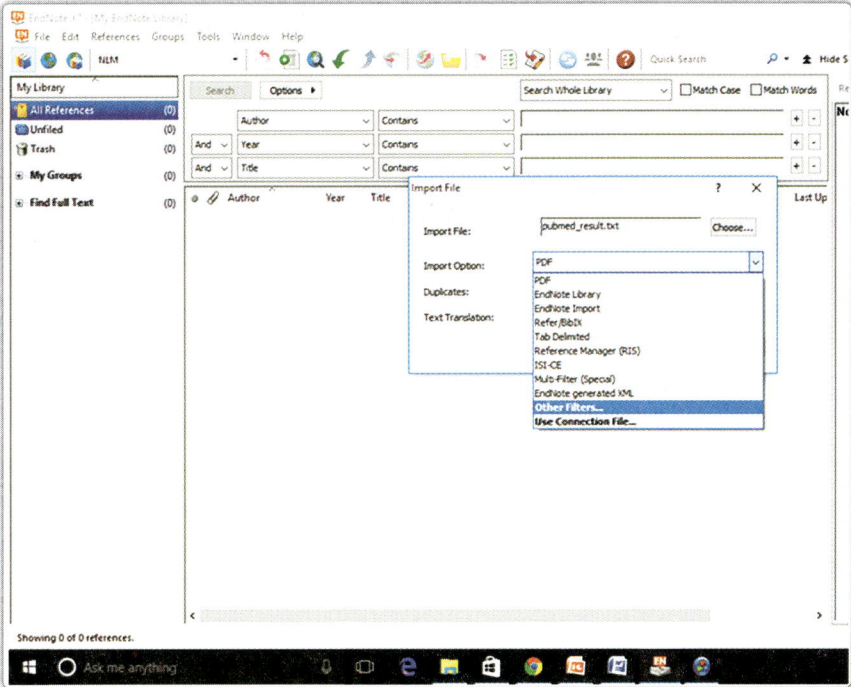

Fig 26. EndNote - Import filters

Source: EndNote X7.3.1 (Thomson Reuters) Recent versions are now with Clarivate Analytics

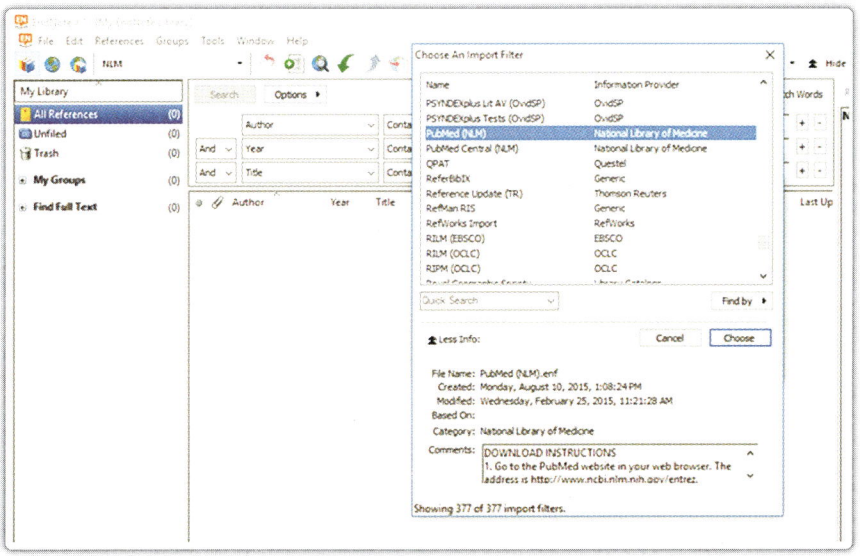

Fig 27. EndNote - PubMed Import filter

Source: EndNote X7.3.1 (Thomson Reuters) Recent versions are now with Clarivate Analytics

- Click **Import**. The references will appear in your active EndNote Library.

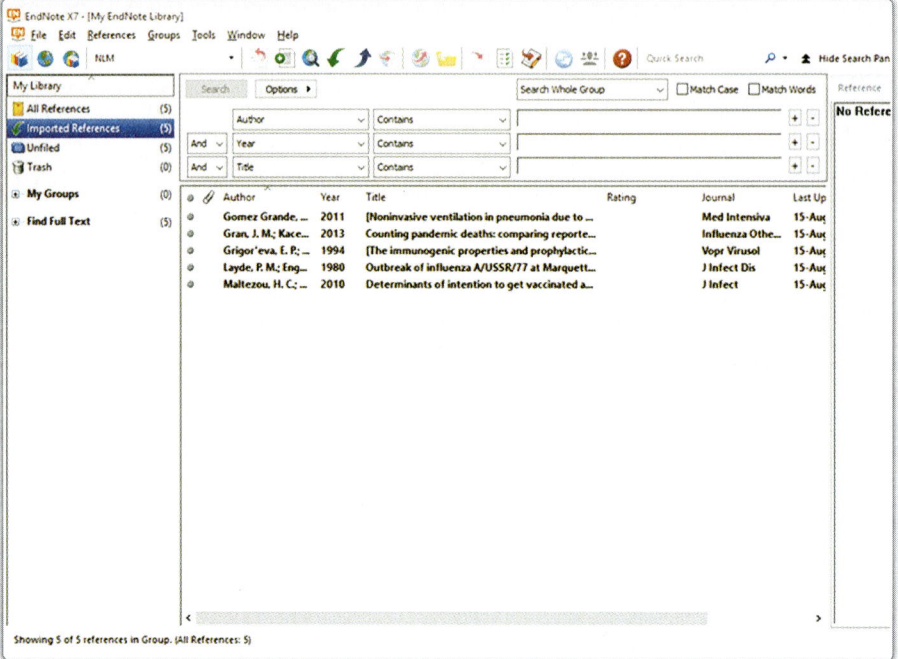

Fig 28. EndNote - Imported PubMed references in my library

Source: EndNote X7.3.1 (Thomson Reuters) Recent versions are now with Clarivate Analytics

Method 2: Search Library of Congress / PubMed / Web of Science /ProQuest / EBSCO / Gale /OvidSP and many more directly in EndNote

- **Search PubMed and other database (**Freely Available Databases and Authentication based**) directly into EndNote.** This Method good for **Known** Citations
- In an EndNote Library, under **Online Search** on the left, click on **PubMed (NLM).**

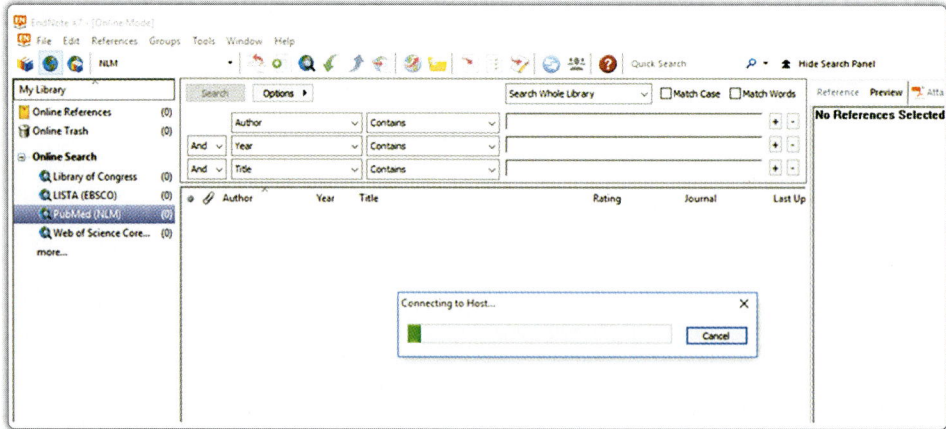

Fig 29. EndNote - PubMed search windows within EndNote

Source: EndNote X7.3.1 (Thomson Reuters) Recent versions are now with Clarivate Analytics

- Enter keyword or others in search box as per choice and hit

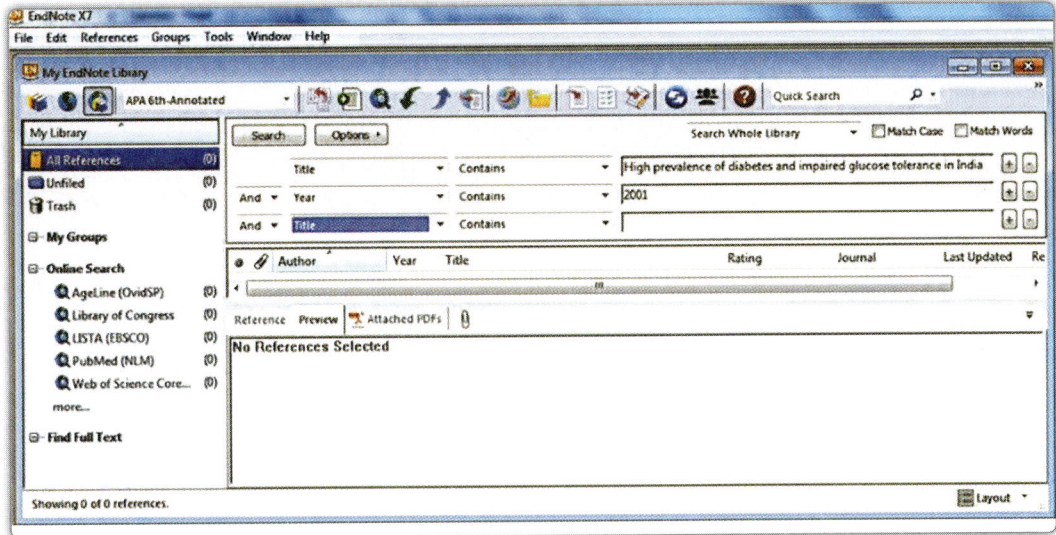

Fig 30. EndNote - PubMed search fields windows within EndNote

Source: EndNote X7.3.1 (Thomson Reuters) Recent versions are now with Clarivate Analytics

- *Retrieved records from* **1** *through* **XX**. Enter the desired number for XX.

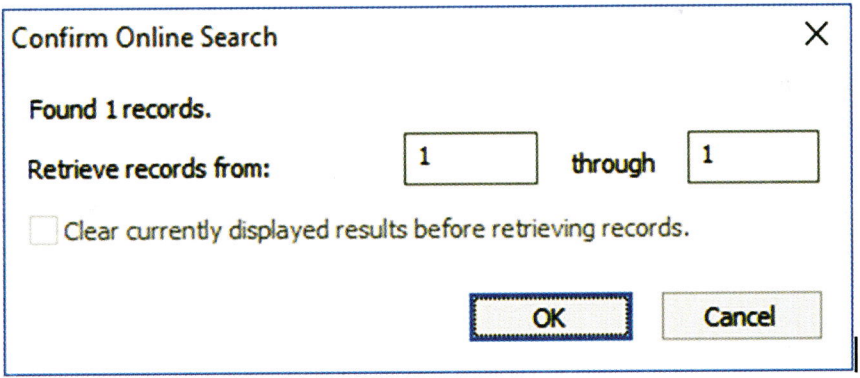

Fig 31. EndNote - PubMed search result windows within EndNote

● All records (relevant?) will be saved into EndNote

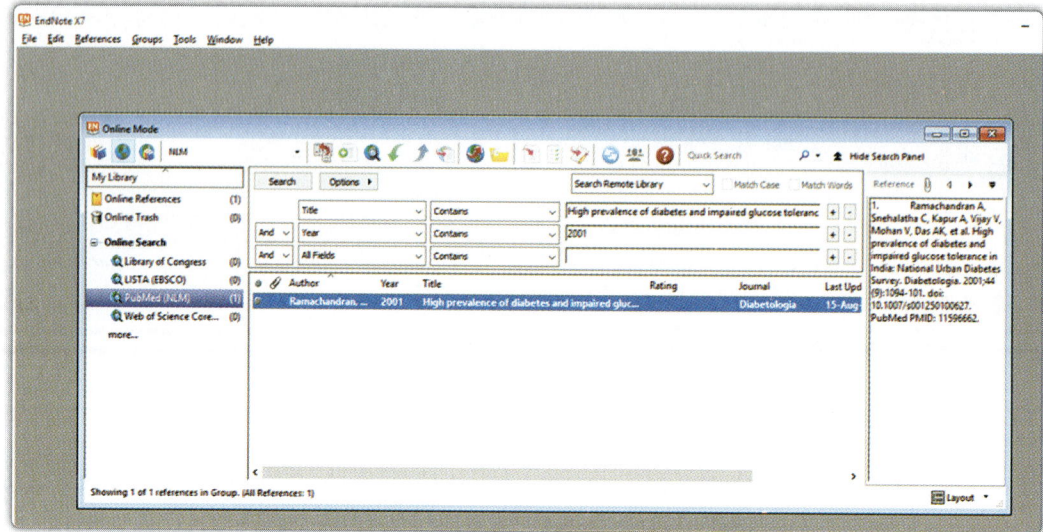

Fig 32. EndNote - PubMed search result in my library

Source: EndNote X7.3.1 (Thomson Reuters) Recent versions are now with Clarivate Analytics

Method 3: Manually Create a Reference in EndNote

● In an EndNote Library, at the top, click on References → New Reference

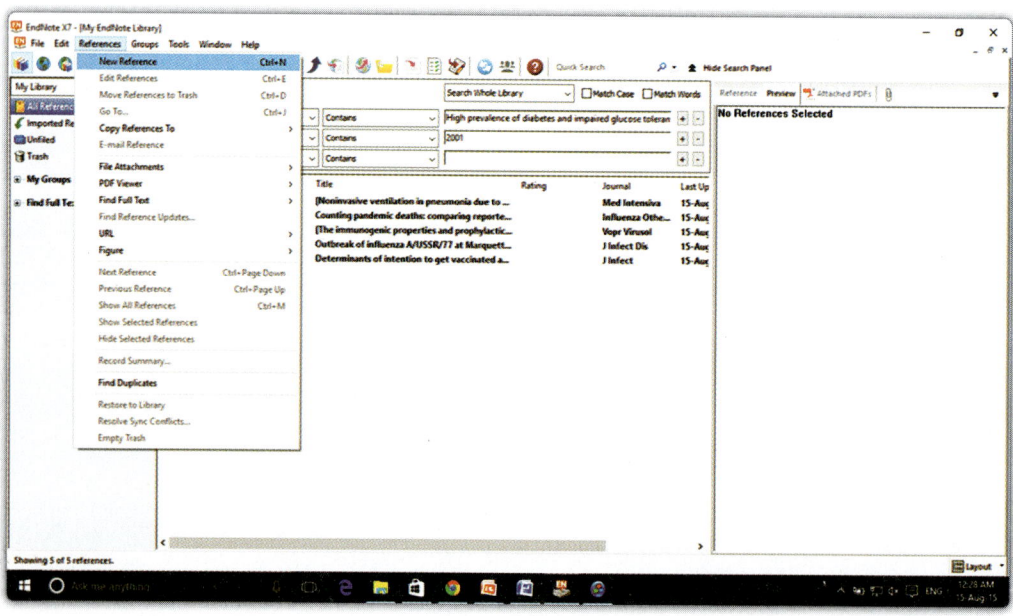

Fig 33. To create a reference in EndNote

Source: EndNote X7.3.1 (Thomson Reuters) Recent versions are now with Clarivate Analytics

- Under Reference Type:, click on the drop-down menu and select accordingly (eg. Journal Article. Book or Web Page)

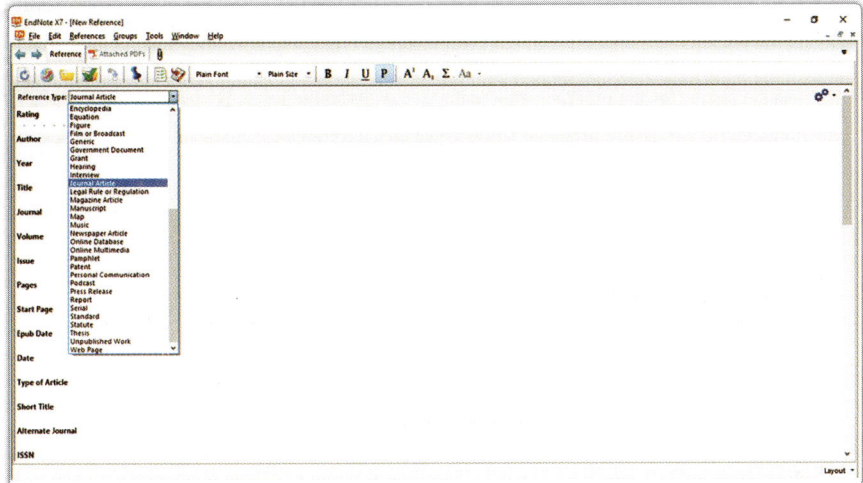

Fig 34. EndNote - Manual reference entry

Source: EndNote X7.3.1 (Thomson Reuters) Recent versions are now with Clarivate Analytics

- **Enter information such as *author, year, title***

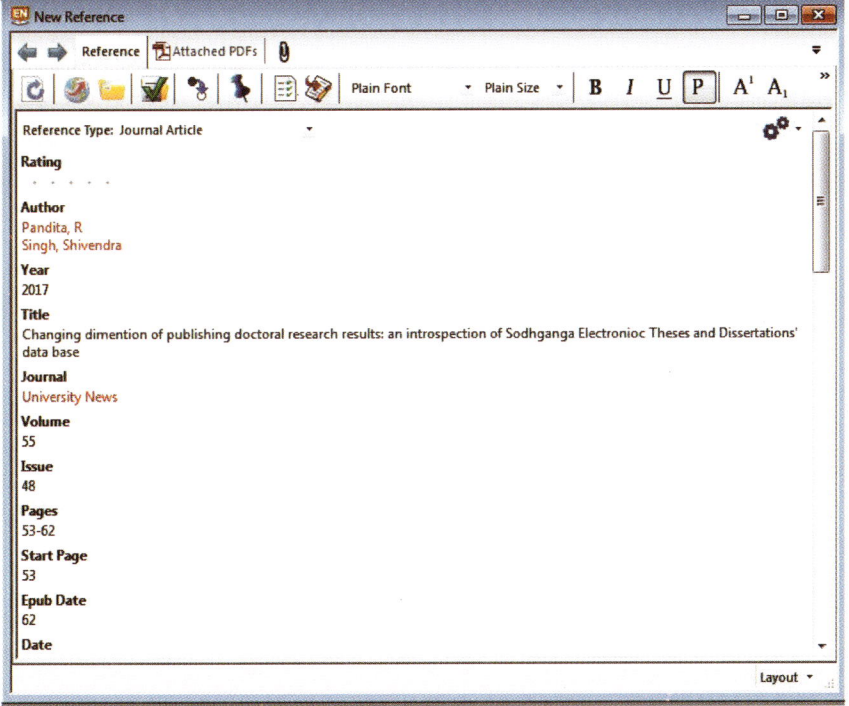

Fig 35. EndNote-Manual reference entry

Source: EndNote X7.3.1 (Thomson Reuters) Recent versions are now with Clarivate Analytics

- Simply click on X (close the window) and the reference will be saved

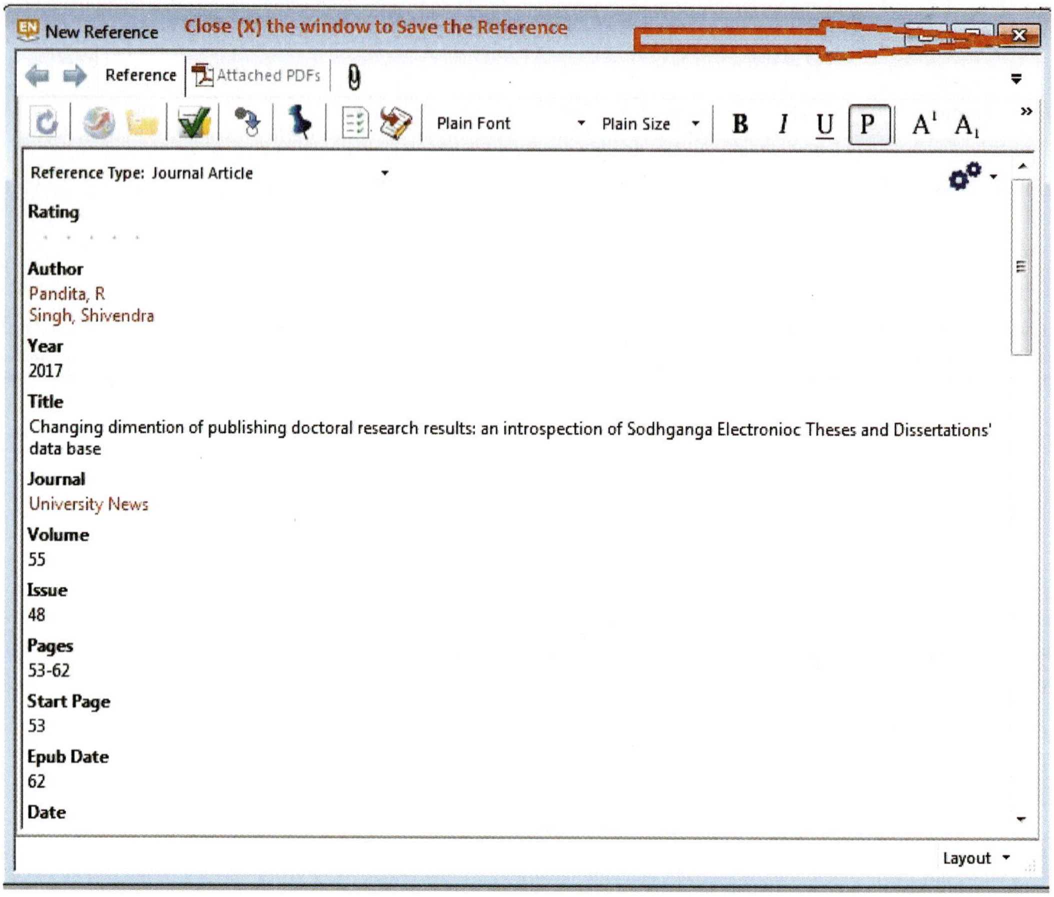

Fig 36. EndNote-Save the manual reference entry

Source: EndNote X7.3.1 (Thomson Reuters) Recent versions are now with Clarivate Analytics

OTHER FEATURES

Removes Duplicate References

- In an EndNote Library, click References on the menu bar → Find Duplicates
- Organize References into Groups (or folders)
  - Highlight a reference in EndNote Library
  - Right – Click → Add References To → Create Custom Group…
  - Type a group name. Enter.
  - Right – Click → Copy References To → New Library
  - Also Cut and Copy
- Auto-Filtering with Smart Groups
- Compress a Library and Email to Colleague

EDIT A REFERENCE

Select the reference and double click. It will open all fields of the reference in editable mode.

File Attachments: You can attach files or insert links to files on your network or hard drive by selecting **File Attachments**, then **Attach File**, from the **References** menu.

Figure: This is another way of attaching a file. To insert a file, select **References** > **Figure** > **Attach Figure,** then click on the **Choose File** button and locate the file you want to attach in the Figure field

EXPORT AND IMPORT OF REFERENCES

The exact method for using direct export will vary with the data provider you are using. A sample of the data providers that support direct export for at least some of their databases includes: Web of Knowledge (information on the Web of Knowledge training options can be found at http://science.thomsonreuters.com/training/wok/), All importing uses a pattern matching process where the pattern of the tags in the data is matched against the pattern of the tags in the import filter.

CITE WHILE YOU WRITE IN MS WORD

Open the EndNote library or libraries that contain the references you wish to cite. Highlight a reference in your EndNote Library by clicking on it once (multiple references can be selected by holding down the Ctrl key) Start Microsoft Word and open the paper you are writing. When you are ready to cite a source, position the cursor in the text where you would like to put the citation. This 🖱 button will insert the references you selected in EndNote into your document at the location of the Word cursor.

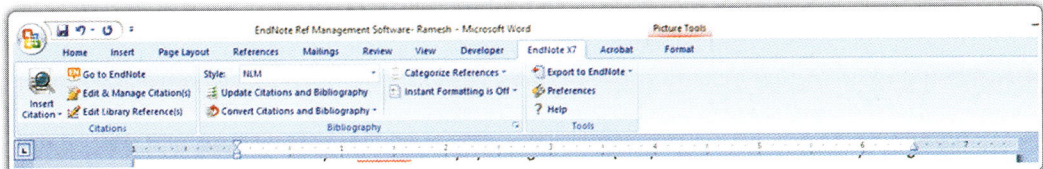

Fig 37. EndNote Tool Bar in MS Word 2007

Insert References in Word

a. In an EndNote library, highlight a reference.

b. In Microsoft Word, place cursor at insertion point.

c. Look for the EndNote toolbar:

   i.   For Word **2007**, click on "EndNote X7" Tab on the Ribbon

   ii. Click on Arrow point in Insert Citation

   iii. Click on Insert Selected Citation(s)

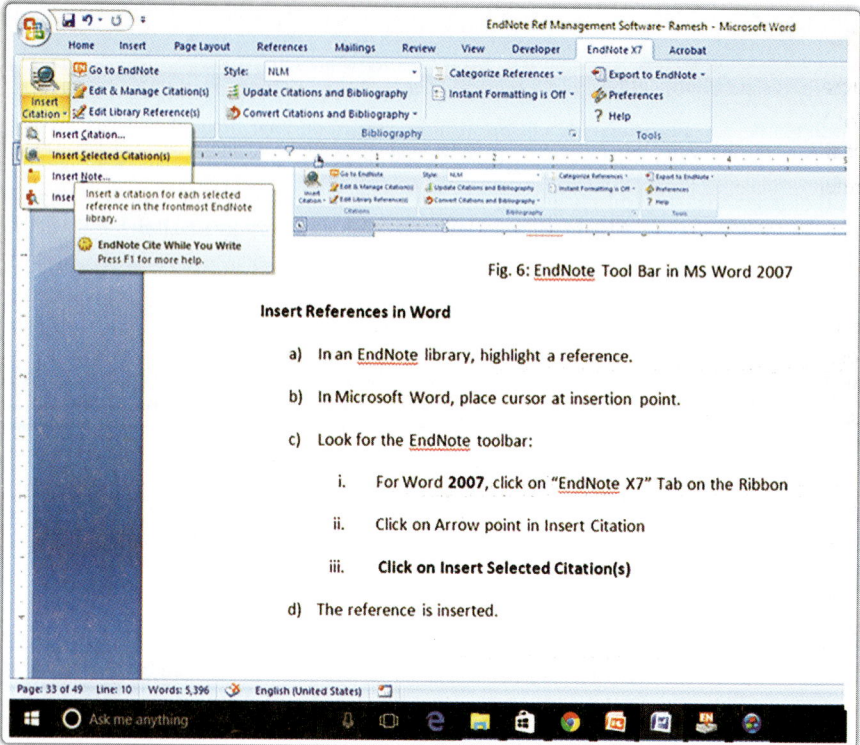

Fig 38. MS Word 2007 - Insert a citation

- The reference is inserted/citing and reference Style

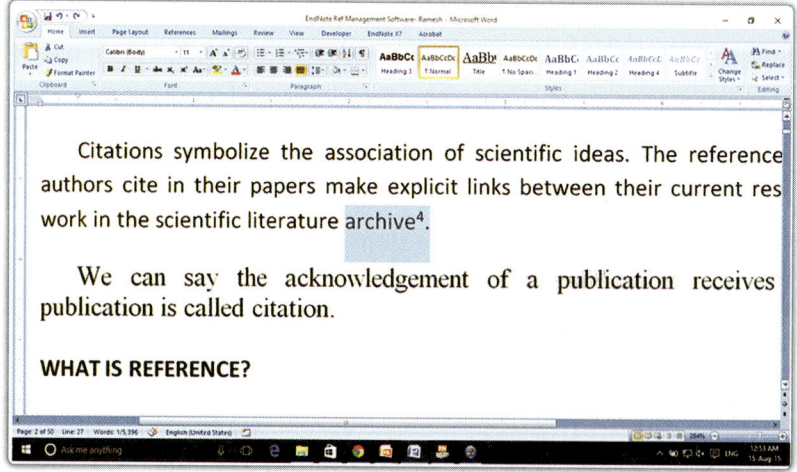

Fig 39. MS Word 2007 - View in citation style on the text

Change Reference Style

a.   In Microsoft Word, look for the EndNote toolbar:

    i.   For Word 2007, go to **Style: → Select another style**

b.   **Select the desired journal style e.g. ACS and Click OK**

c.   The references are now re-formatted.

CONCLUSION

Reference management software have come as a handy tool for researchers to cite different sources of information, from which they seek the information related to their study. These reference management software are helpful in citing sources both within the text and as acknowledgement under references. Reference management software reduce the citing and referencing errors and are quite useful in avoiding duplication of references. Reference management software are equally useful in choosing the citation style. Author must be aware about the journal reference style. Matching reference style with a particular research journal is always a prerequisite, before submitting an article or ever before sending the article for peer review. It is a very comment, which most of the researcher receives from the editors that reference style is not matching as per the journal standard. Accurate reference style has its own impact on impact factor of the journal. The EndNote software helps authors to capture offline and online references, in organizing the references, in generating the bibliography and most importantly formatting the reference style.

BIBLIOGRAPHY

1. American Psychological Association, Publication manual of the American Psychological Association. 6th ed.; American Psychological Association: Washington, ; 2009; p 272.

2. Answers Corporation, : What is the difference between a bibliography and reference? [online] Available from http://wiki.answers. com/ Q/ What_is_the_ difference_ between_a_bibliography_and_reference ([accessed Accessed 06th November 6 2010).].

3. Caprette, D. R., Common Errors in Student Research Papers. [online] Available from http://www.ruf. rice.edu/~bioslabs/tools/ report/reporterror.html ([accessed Accessed 08 November 8, 2010).].

4. Clarivate Analytics. The Concept of Citation Indexing: A Unique and Innovative Tool For Navigating the Research Literature [online] Available from https://clarivate.com/essays/concept-citation-indexing/

5. Division of Teaching & and Learning Services- Central Queensland University, Harvard (author-date) referencing guide. University, C. Q., Ed. Central Queensland University: 2007; p 49.

6. Garfield, E., When to cite. Library Quarterly. 1996, ;66, :449-58.

7. Gibaldi, J., MLA Handbook for Writers of Research Papers. 6th ed.; 2003.

8. Harris, R. A., The Plagiarism Handbook: Strategies for Preventing, Detecting and Dealing with Plagiarism. Pyrczak: Los Angeles, 2001; p 208.

9. International Committee of Medical Journal Editors Uniform Requirements for Manuscripts Submitted to Biomedical Journals (Vancouver System). [online] Available from http://bioclima.ro/urm_full.pdf.

10. Middelberg, A., How to review a paper? How to Review review a Paperpaper-Specific specific Comments comments on Chemical chemical Engineering engineering Sciencescience? In Elsevier Reviewer Workshop, Elsevier Tianjin University, 2010.

11. Modern Language Association, MLA Handbook for Writers of Research Papers. 7th ed.; New York: MLA; New Yor, 2009; p 292.

12. Thomson Reuters (ISI) EndNote Class Outline: Building Your EndNote Library. [online] Available from www.endnote.com/ training/EndNote_X4_Class_Outlines.pdf ([accessed Accessed 08 November 8, 2010).].

13. Thomson Reuters. EndNote Information. [online] Available from http://www.endnote.com/eninfo.asp ([accessed Accessed 07th November 7, 2010).].

14. University of Chicago, The Chigago Manual of Style. 16th ed.; Chicago: The University of Chigago: Chicago,; 2010; . p 1026.

15. What is Citation? [online] www.plagiarism.org/ resources /documentation /.../ what_is_citation.doc ([accessed Accessed 06 November 6, 2010).].

16. Wikipedia Comparison of reference management software. [online] http://en.wikipedia.org/w/index. php?title=Comparison_of_reference_management_software ([accessed Accessed 14 august August 14, 2015).].

17. Wikipedia Reference management software. [online] Available from http://en.wikipedia.org/wiki/ Reference_management_ software (accessed Accessed 07 November 7, 2010).].

Chapter 6

Reference Management Software Mendeley

INTRODUCTION

Mendeley, like EndNote is a reference management software, very popular among the researchers all across the world, used for referencing purposes. Mendeley unlike EndNote is non-proprietary software and one can make a very good use of the software without paying for its services. In the present discussion, an attempt has been made to highlight the advantages of Mendeley as reference management software and how far it can be used effectively to cite while write. The discussion also lasts around some added advantages of Mendeley over that of various other proprietary and non-proprietary reference management software.

Mendeley is a free reference manager, social research network, to help the researcher to organize their research reference library. This is the first reference software, which can extract metadata from PDF files like title, authors, journal name, year, volume and issue number. Mendeley creates a private library for every user, accessible at (https://www.elsevier.com/connect/victor-hennings-brief-guide-to-mendeley).

The software founders were three German (Paul Foeckler, Victor Henning and Jan Reichelt) research scholars in 2007. The software is named after the scientist Gregor Mendel and Dmitri Mendeleev. The founders had the idea of combining two researchers' names: Mendeleev and Gregor Mendel and that is how they came up with "Mendeley". (http://www. doctorpreneurs.com/paul-foeckler-interview/) In April 2013, the Mendeley was acquired by the Elsevier.

Paul Foeckler, Victor Henning, Jan Reichelt (From left to right)

VERSION

2007	-	Founded
2008 (August)	-	First Beta version
2016	-	Mendeley Desktop v1.17.6

Research writing is as tiresome as doing research itself. It is always a cumbersome process to keep track of all bibliographical details of the earlier research works consulted, while undertaking a new research activity. Thereon what become more difficult is to cite the different sources of information consulted while undertaking the research and to overcome this problem of citing the earlier works in one's own work.

WHAT DOES MENDELEY DO?

- Imports references from your computer, flash drive or the internet
- Extracts citation info from added pdfs
- Add entry manually
- Add to watch folder
- Organized the references
- Generates citations and bibliographies
- Create folders
- Create groups and share the documents

HOW TO WORK WITH MENDELEY?

The researcher, student or any person who wants to install Mendeley on the computer, then they go to the Mendeley website i.e. HYPERLINK "http://www.mendeley.com" www.mendeley.com (Fig. 1). So first thing go to create an account (Figs 2 to 4), or those already have account to log-in the welcome webpage of Mendeley. After the log-in account on Mendeley you have option to download the Mendeley Desktop client software (Figs 5 to 7). The Desktop client can use for your citations on your paper, so its need to download and install (Figs 9 to 14). After the installation you can see the Mendeley Icon on All Programs menu as well as Desktop.

1. Open Mendeley web site (www.mendeley.com) on the Internet Browser

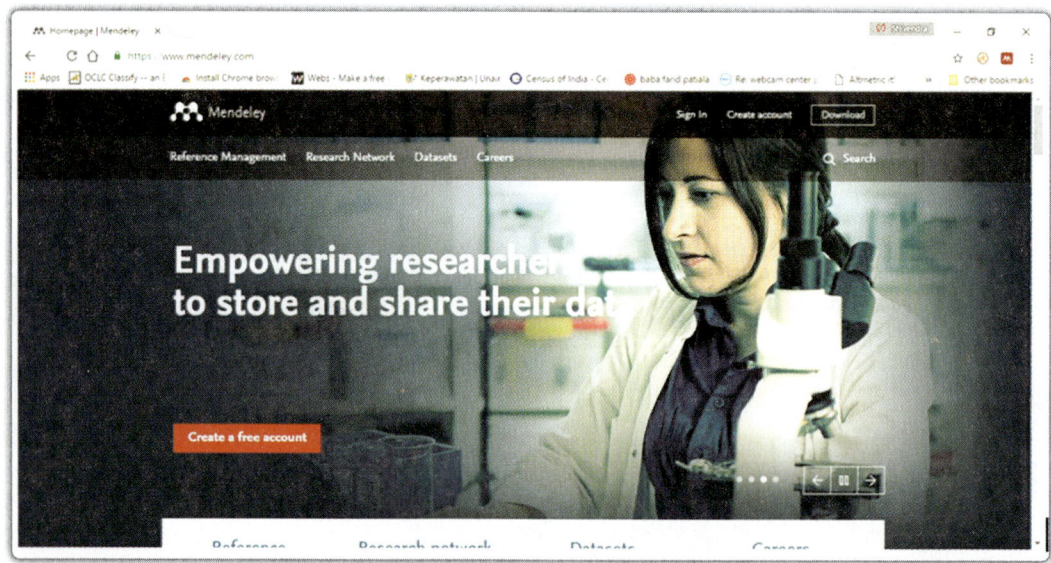

Fig 1. Mendeley Website - Welcome page

Source: Elsevier

2. Create free account to join the mendeley (One account for all your research)
3. To install Mendeley one needs to first sign up for your account on the website (www.mendeley.com)
4. Start from your basic details

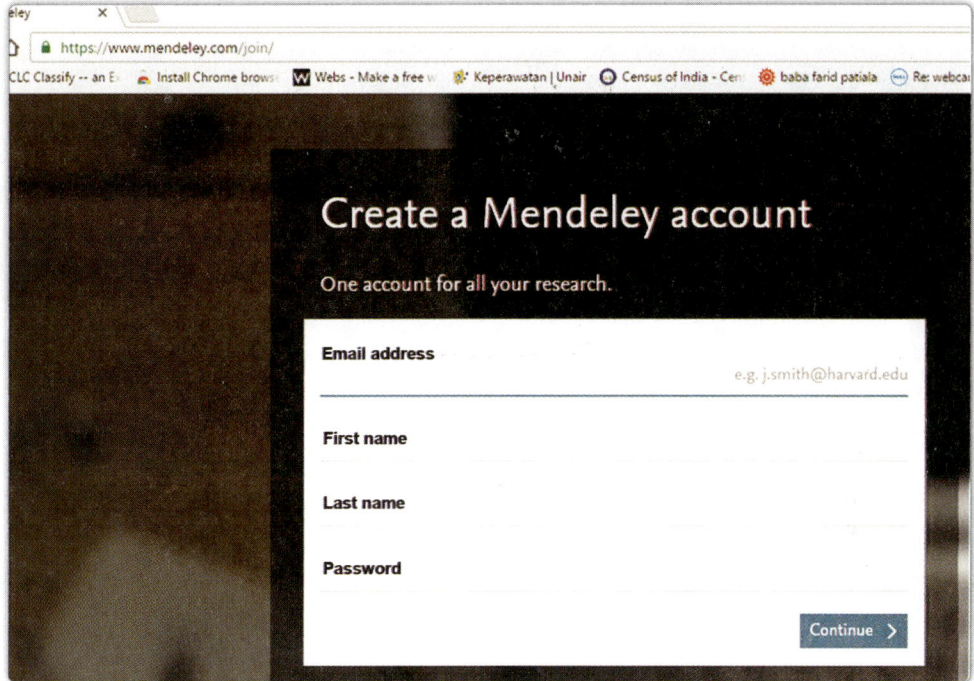

Fig 2. Creating a mendeley account
Source: Elsevier

5. Use Mouse drag to select your subject field and Academic status

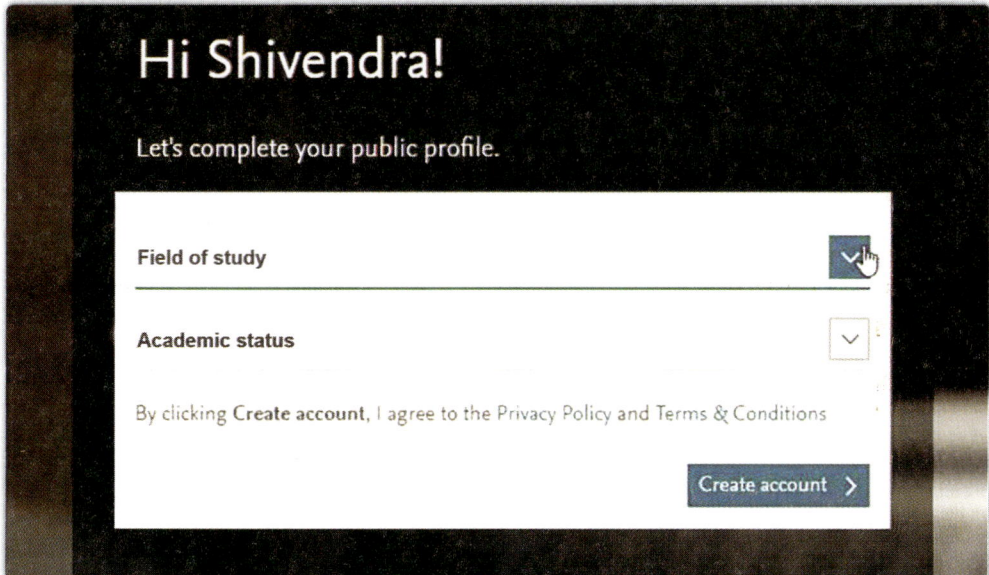

Fig 3. Creating a mendeley account
Source: Elsevier

6. Using keyboard input on "Select an institution (window)" in "Select an institution

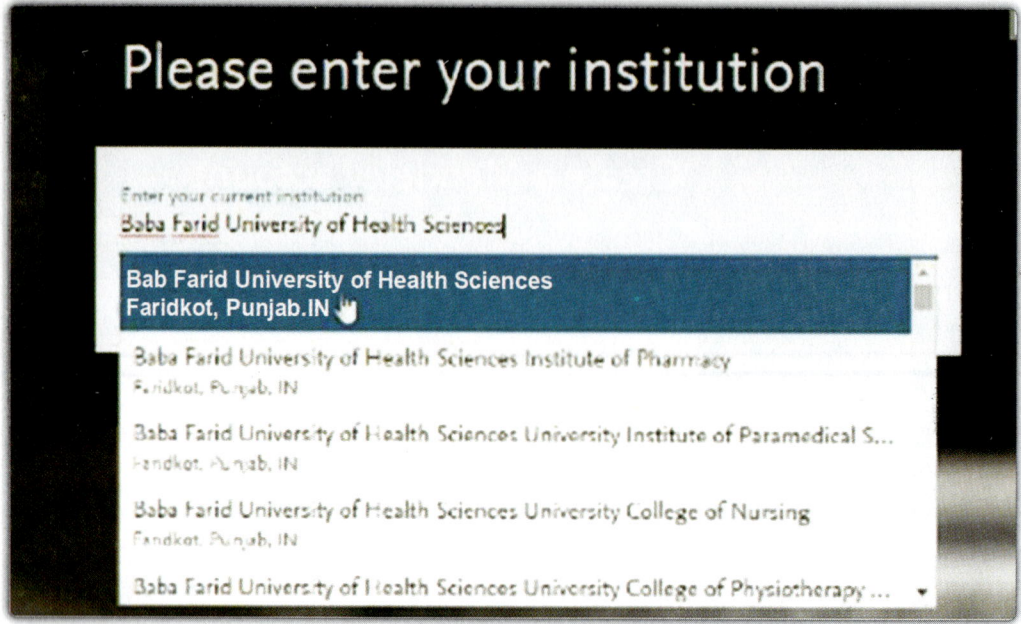

Fig 4. Creating a mendeley account

Source: Elsevier

7. After the Signing up one can download and install the Mendeley Desktop version.

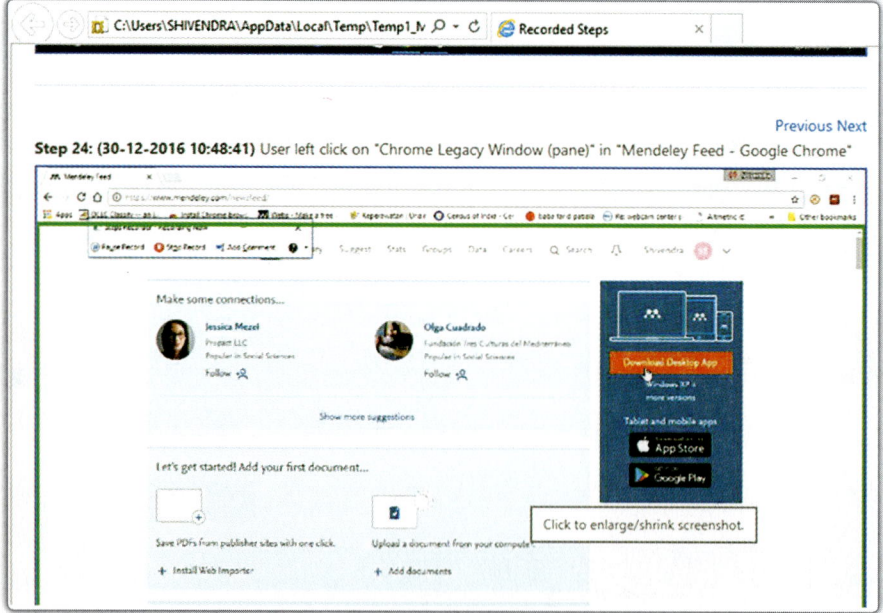

Fig 5. Download mendeley desktop client software

Source: Elsevier

8. If one has forgotten to download at the same time or want to download it later on, then one needs to go to one's personal setting on user account and then click 'Download Mendeley'

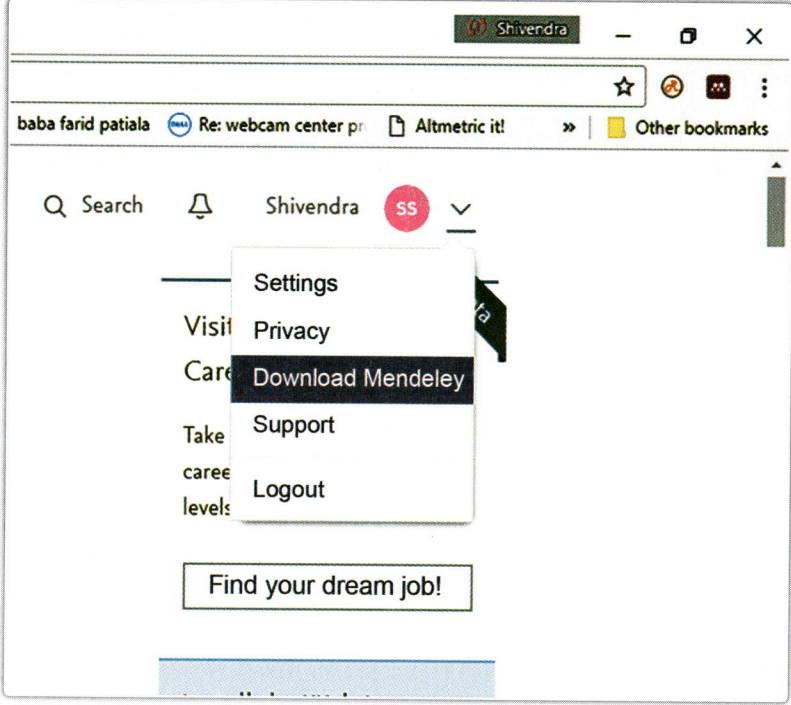

Fig 6. Download mendeley desktop client software

9. Click on 'Download Mendeley' to start downloading

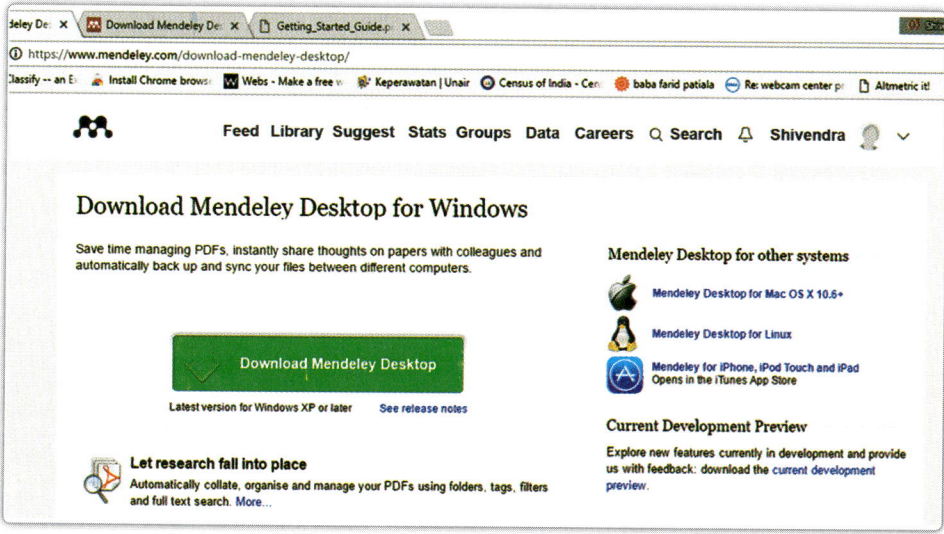

Fig 7. Download mendeley desktop client software

MENDELEY DESKTOP

1. Download Mendeley Desktop (Save Mendeley Desktop on your hard drive.)

Mendeley-Desktop-1.17.6-win32
31-12-2016 00:48
53.5 MB

Fig 8. Mendeley software window installation file

2. Install Mendeley Desktop

Double-click the Mendeley Desktop icon in your browser's download window and follow the instructions on screen. Click on "Run (button)' in open file.

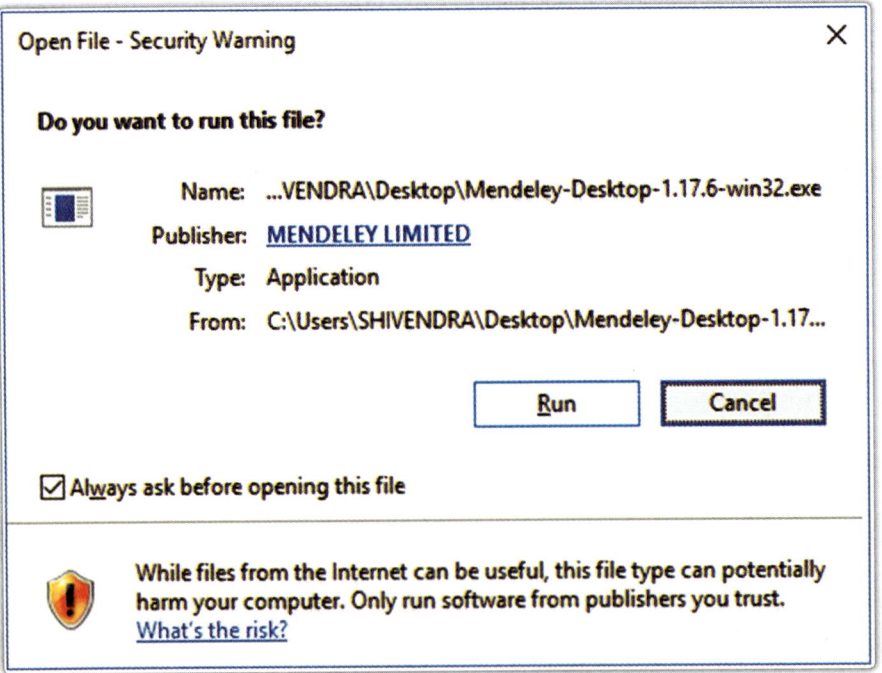

Fig 9. Mendeley desktop installation process

3. Click 'Next'

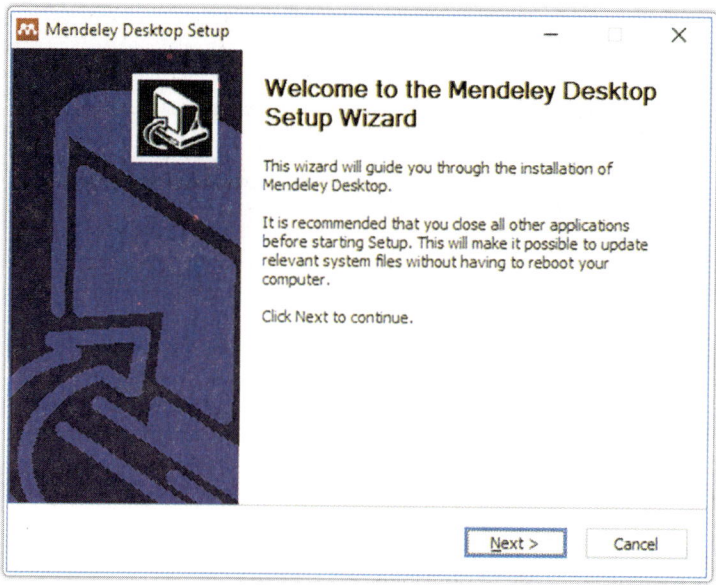

Fig 10. Mendeley desktop installation process

4. Click on "I Agree (button)" in "Mendeley Desktop Setup "

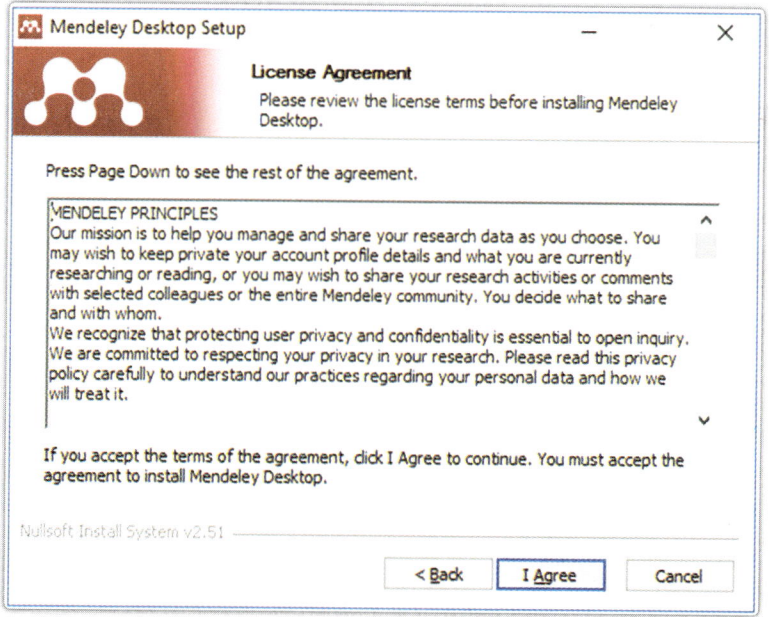

Fig 11. Mendeley desktop installation process

Source: Elsevier

5. Click on "Next > (button)" in "Mendeley Desktop Setup "

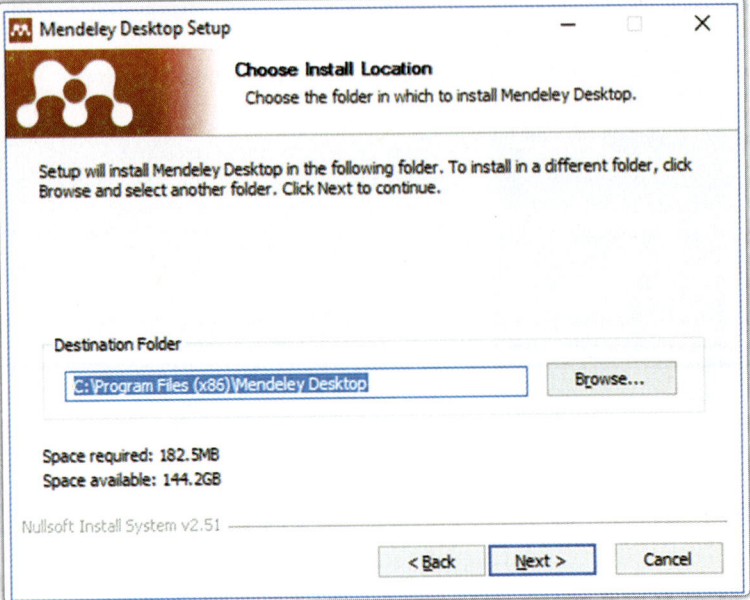

Fig 12. Mendeley desktop installation process

Source: Elsevier

6. Click on 'Install' in "Mendeley Desktop Setup

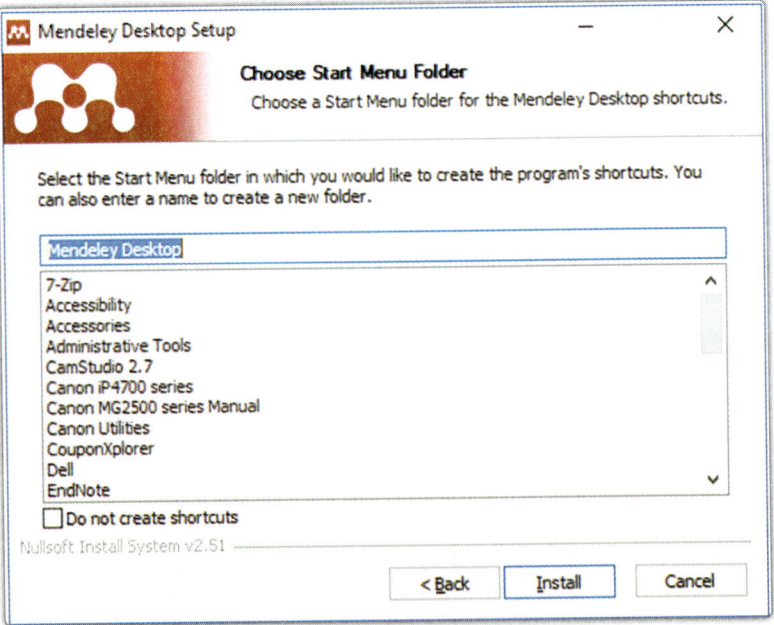

Fig 13. Mendeley desktop installation process

Source: Elsevier

7. The process will be running till final installation

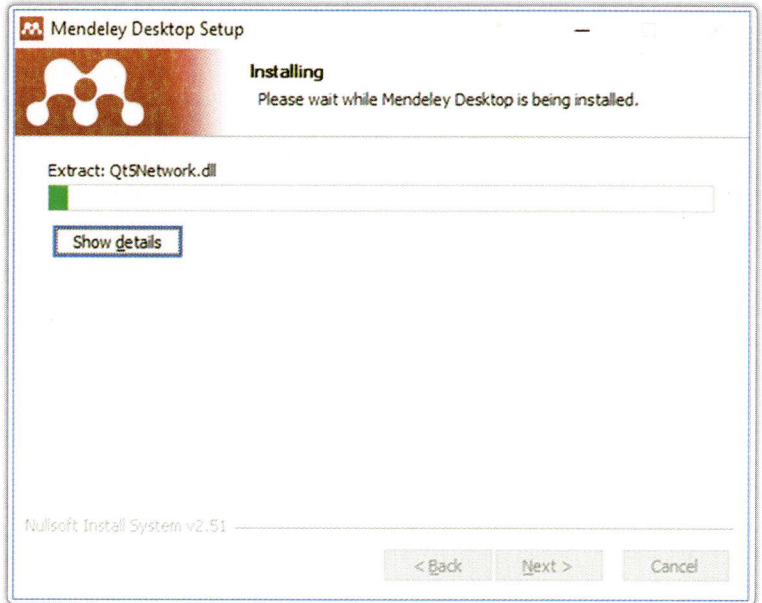

Fig 14. Mendeley installation in process

Source: Elsevier

8. Click on "Finish (button)" in "Mendeley Desktop Setup "

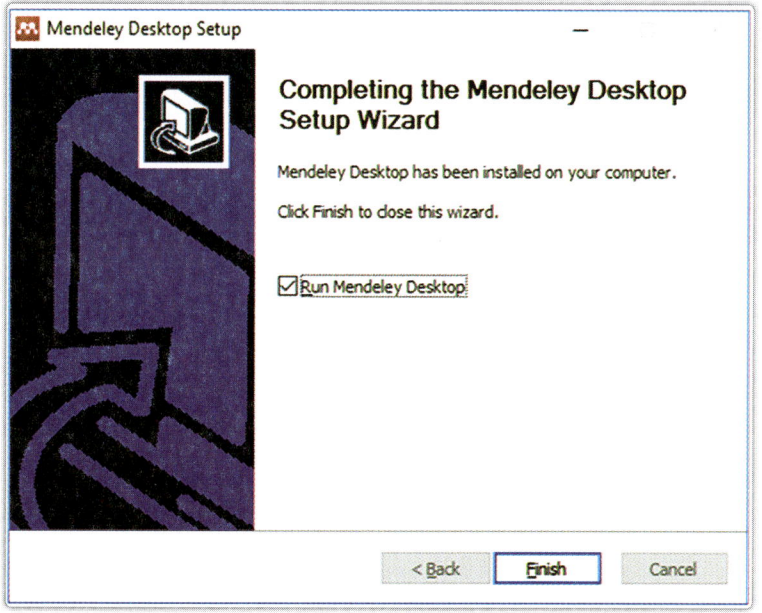

Fig 15. Mendeley desktop installation process

CREATING YOUR LIBRARY IN MENDELEY DESKTOP

There are several ways to create your library in Mendeley like import your existing library, drag and drop PDF file, add a PDF, add folder, through watch folder, add entry manually, through web importer, import file from other reference management software, and add from Mendeley website also. A step by step procedures given below.

Import you Existing Library

1. First to sign in your account

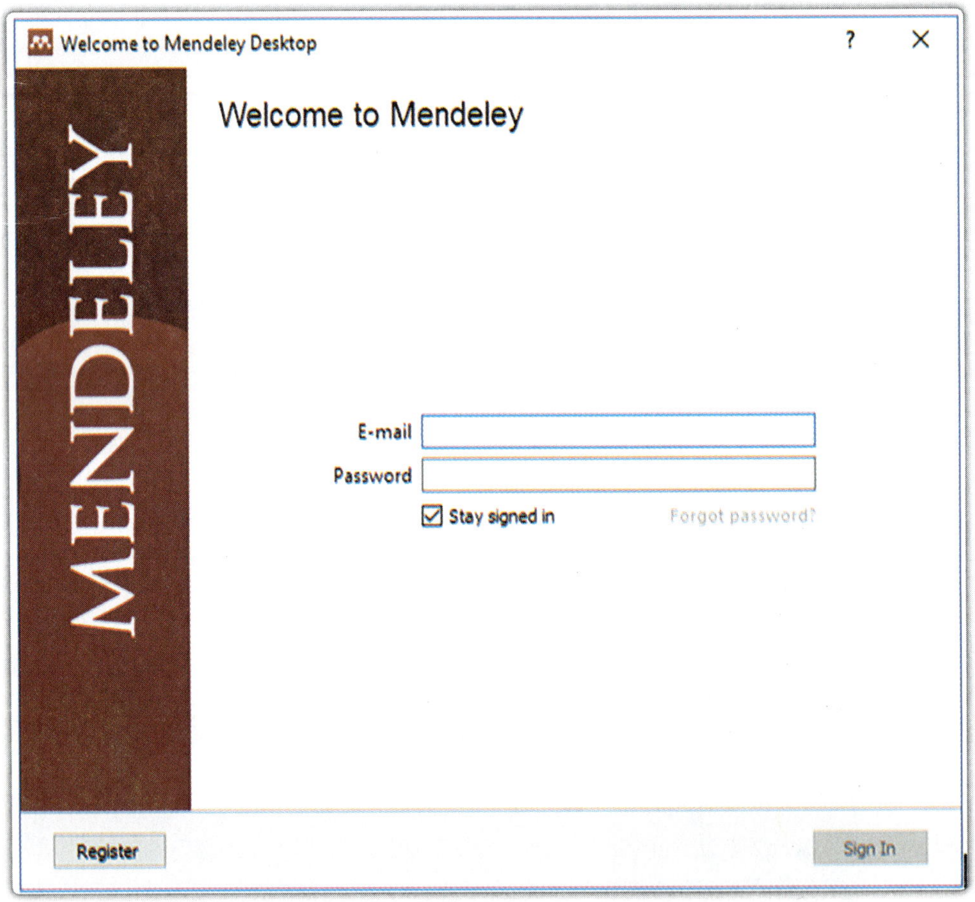

Fig 16. Mendeley desktop login

Source: Elsevier

2. If you have the EndNote file, then say 'Yes" otherwise say 'No thanks". Here I say 'No thanks".

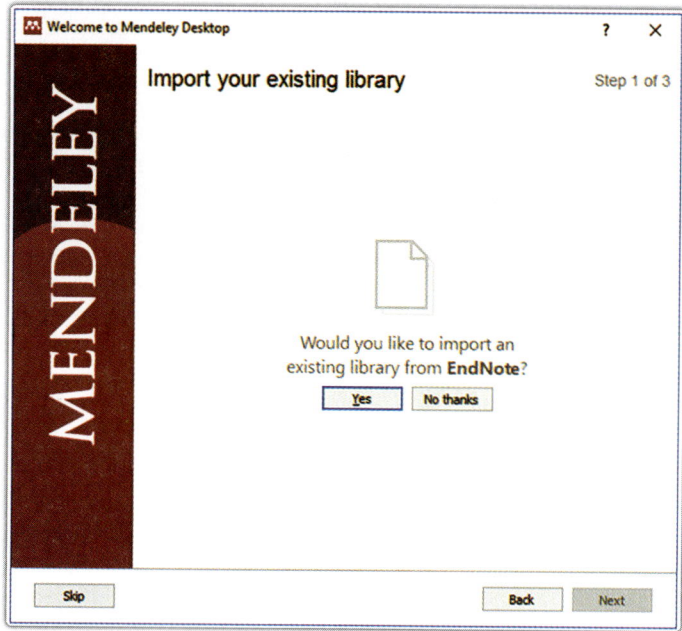

Fig 17. Mendeley desktop - Import an existing library-EndNote library

Source: Elsevier

3. Import your existing library

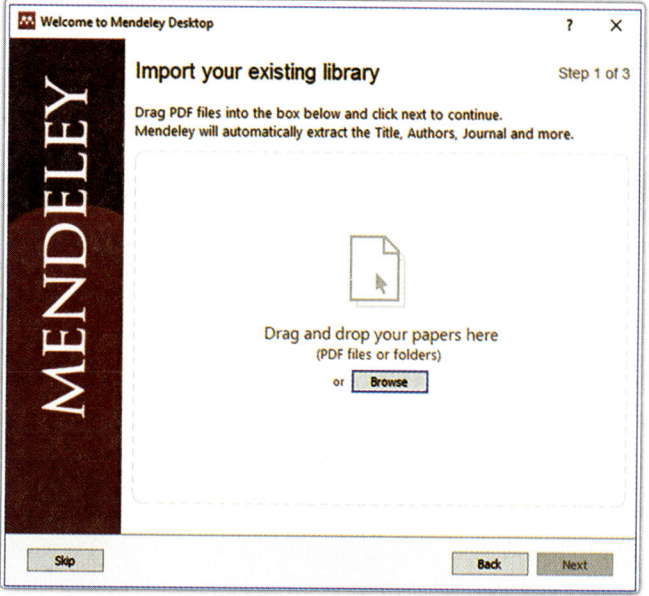

Fig 18. Mendeley desktop-Import an existing library-Pdf files

Source: Elsevier

4. Add PDFs to Mendeley

You can add PDFs into Mendeley by drag and drop pdf files into the box or use browse button to select the pdf files and click next to continue. Mendeley will automatically extract the Titles, Authors, Name of the Journal, Year, Volume Number and Issues etc.

5. User mouse to start to "Welcome to Mendeley Desktop (window)"

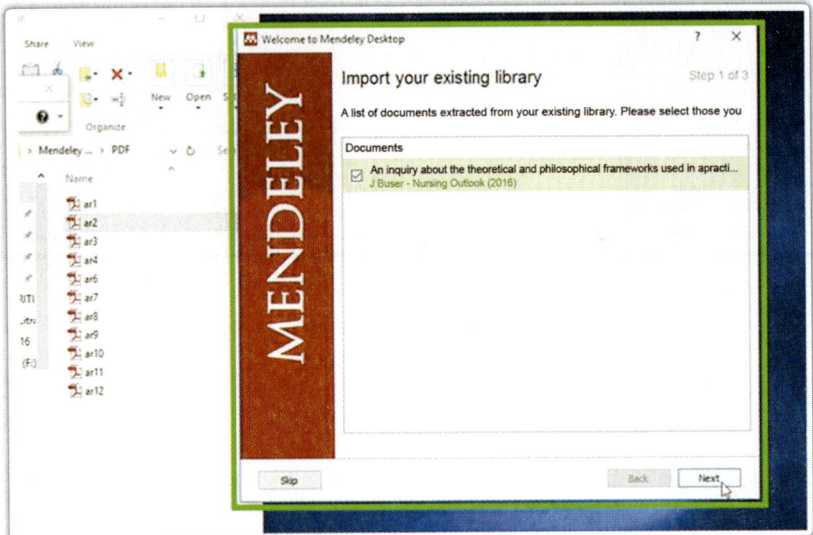

Fig 19. Mendeley desktop-Import an existing library-Pdf files

Source: Elsevier

6. Document details lookup- After finishing the process, you may be able to see the document details on the right side of the window, when one clicks over the document

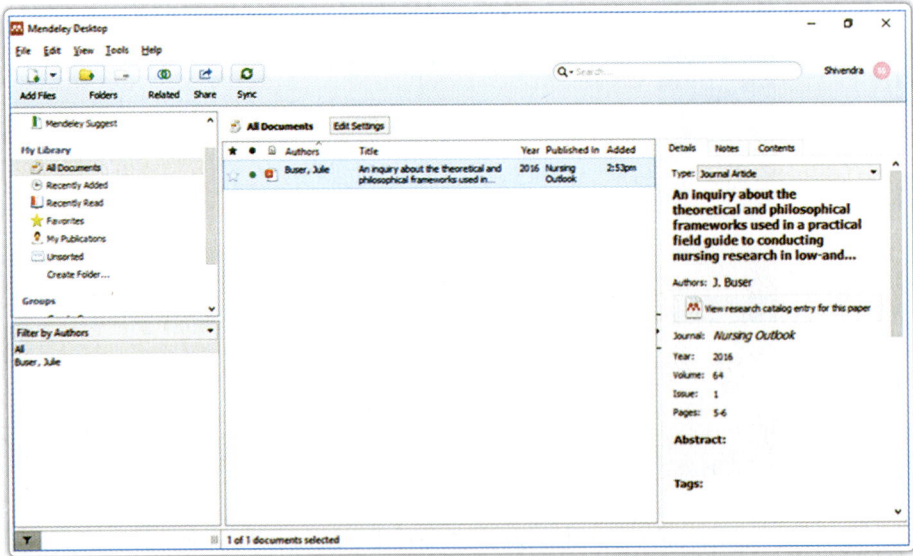

Fig 20. Mendeley desktop-Imported PDF file's Bibliographic details

Source: Elsevier

Add Pdf Files from Windows Menu Bar

1. User left click on "Add Files (button)" in "Mendeley Desktop"

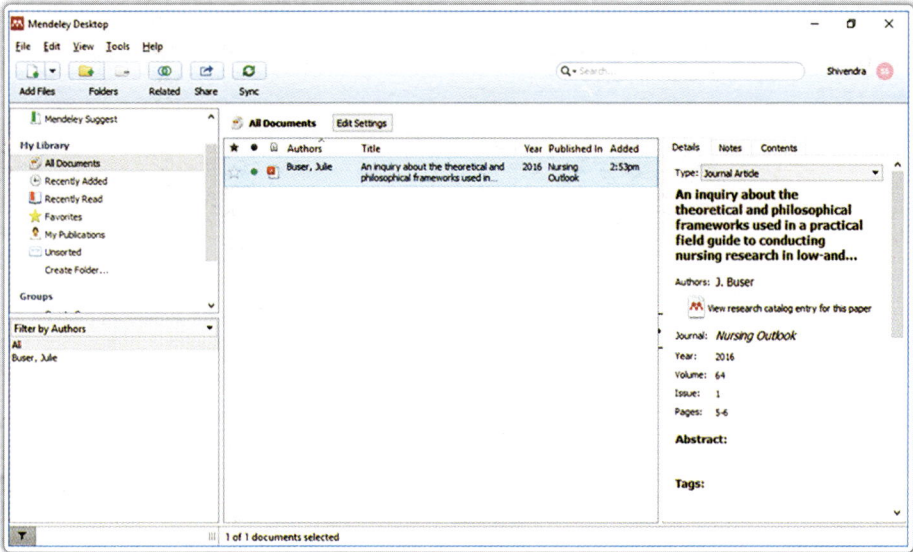

Fig 21. Mendeley desktop-Add pdf files from the desktop menu

Source: Elsevier

2. After choosing from the menu in "Add Files", browse to the location of the stored files, then select the desired file and click the open button.

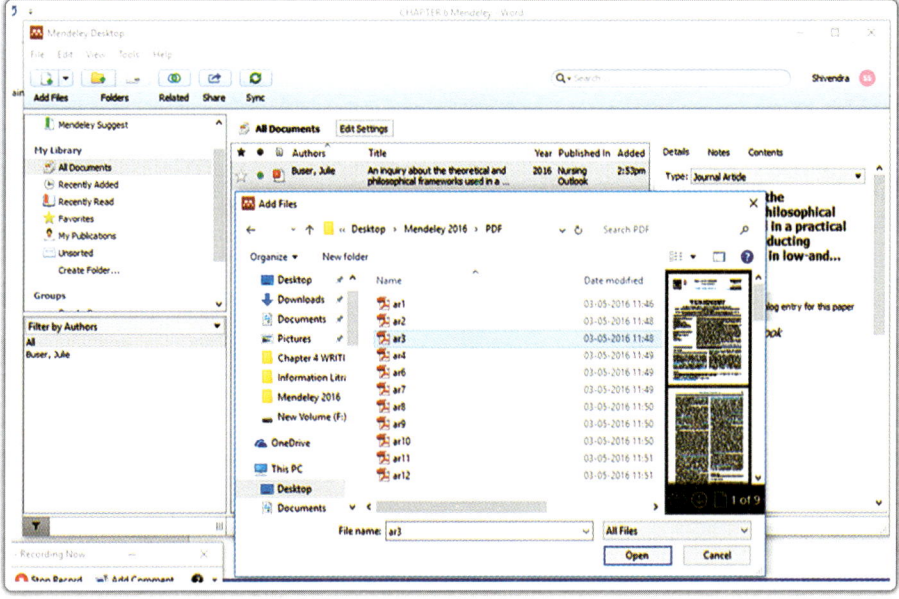

Fig 22. Mendeley desktop-Add pdf files from the desktop menu

Source: Elsevier

3. Now the PDF file has been added to your Mendeley library. The software will attempt to automatically detect the bibliographic data which you can see in the Reference details section on the right side of the window.

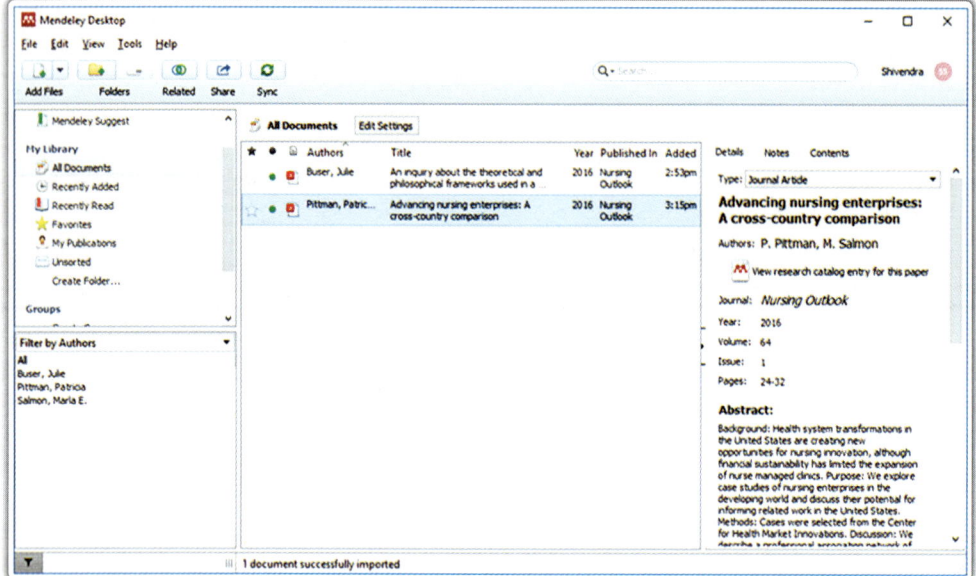

Fig 23. Mendeley desktop-Added pdf file's Bibliographic details

Source: Elsevier

Add Folder

1. Go to Mendeley → File → Add Folder

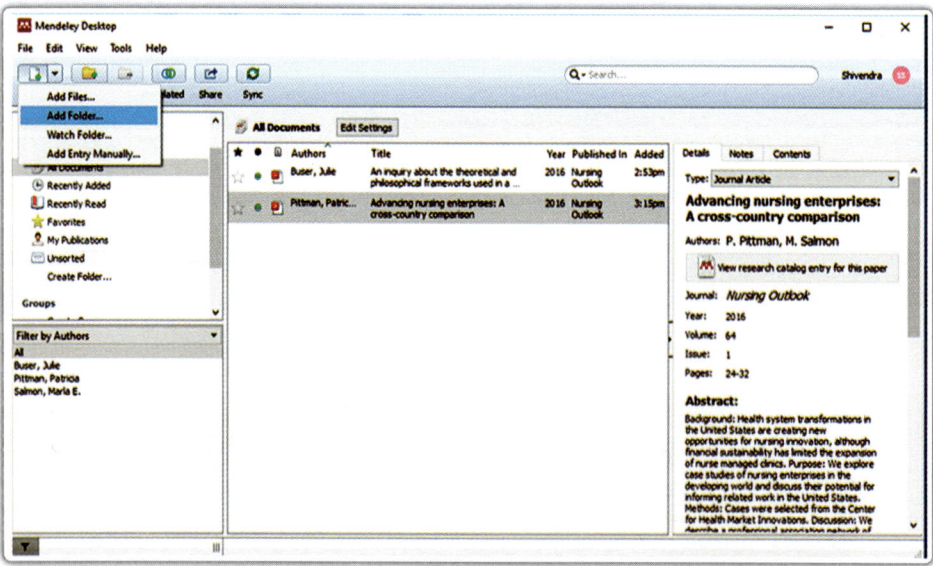

Fig 24. Mendeley desktop-Add document folder

Source: Elsevier

2. Click on Add Folder from Menu Bar and Browse to the location of the PDF file folder and then click the 'Ok' button.

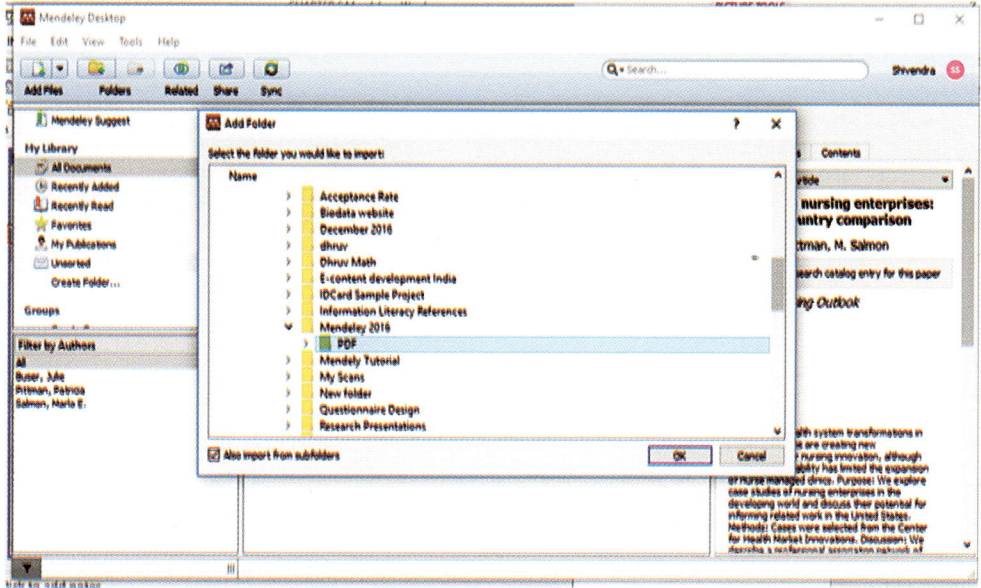

Fig 25.   Mendeley desktop-Selection of document folder

Source: Elsevier

3. All PDFs have been added to the library and their bibliographic details detected automatically.

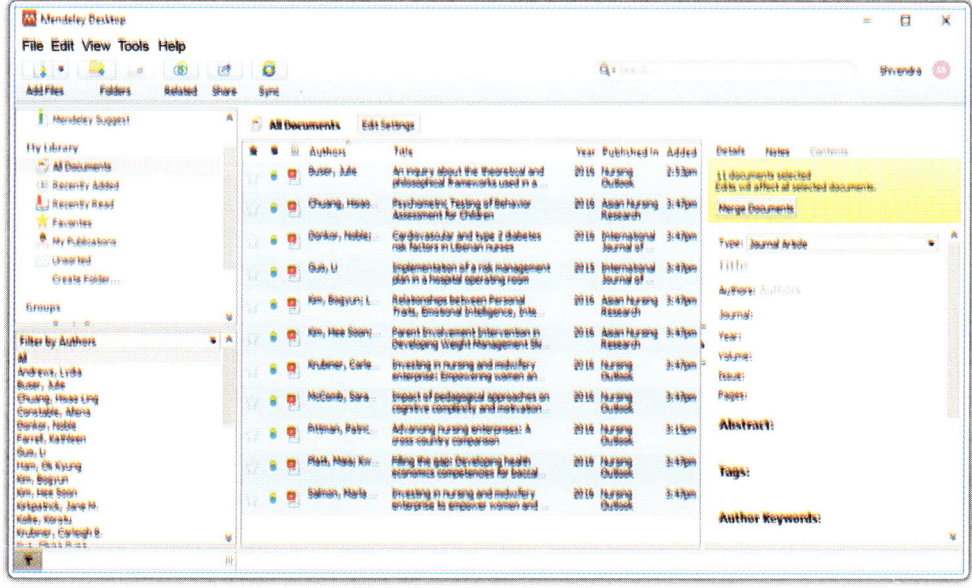

Fig 26.   Mendeley desktop - Documents added from the folder

Source: Elsevier

4. To select the document to see the bibliographical details

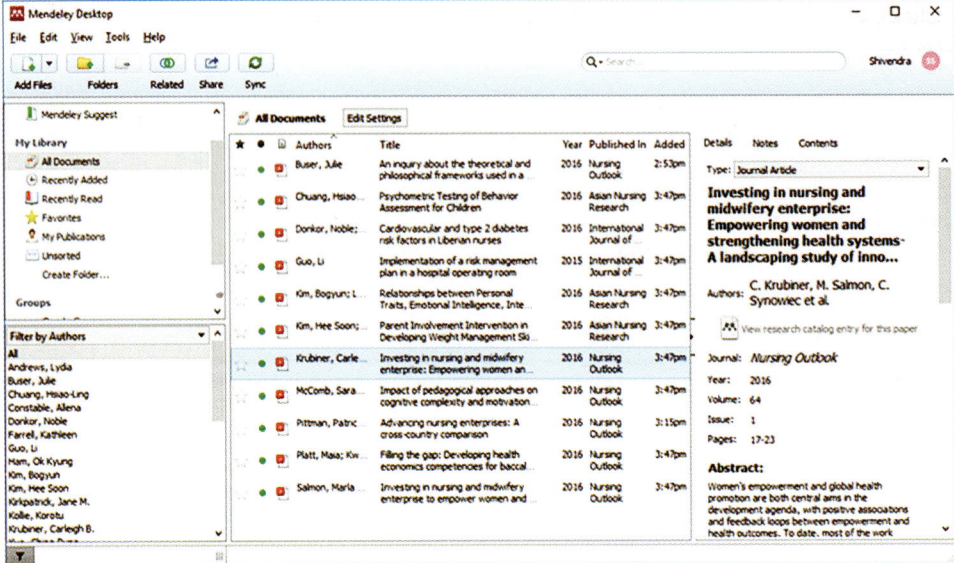

Fig 27. Mendeley desktop-Bibliographic details of a document added from the folder

Source: Elsevier

Add File Through Watch Folders

When we have our PDFs in a specific folder in which we add files regularly, the Mendeley will automatically add these PDFs when we select the "Watch Folder".

Go to Mendeley → File → Watch Folder

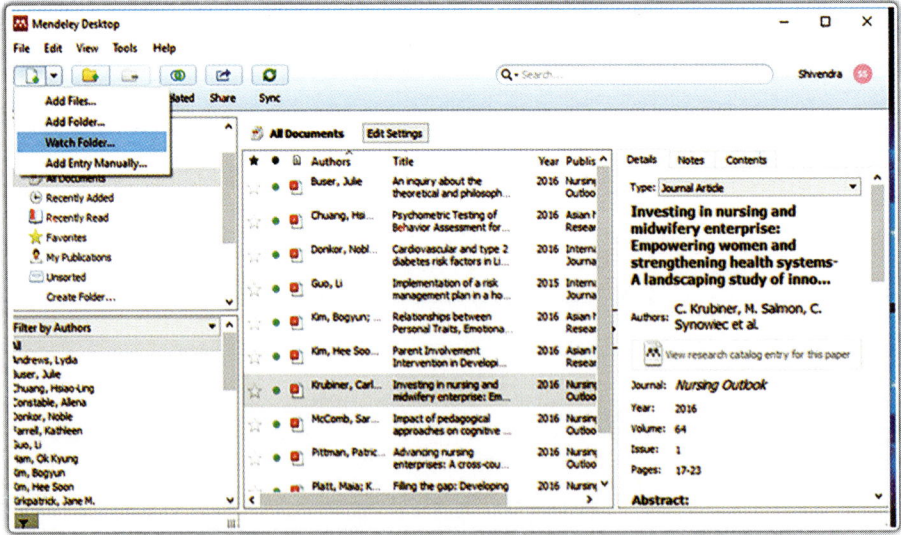

Fig 28. Mendeley Desktop-Watch folder

Source: Elsevier

Add Entry Manually

1. Click on "Add Entry Manually" under the "File" on the menu bar

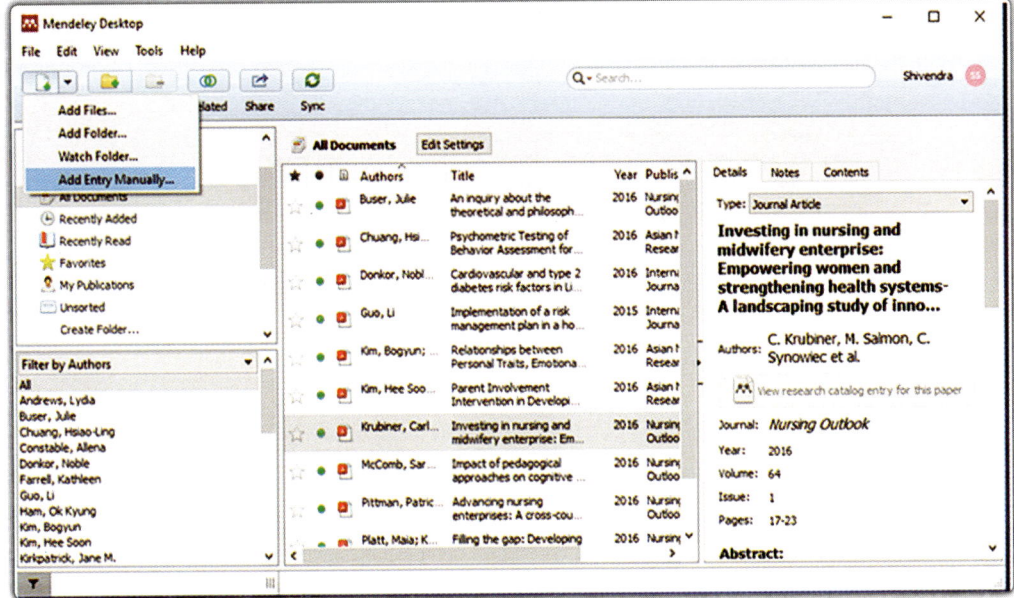

Fig 29. Mendeley desktop-Add entry manually

Source: Elsevier

2. A pop-up window will appear where you can add references, bibliographic details manually.

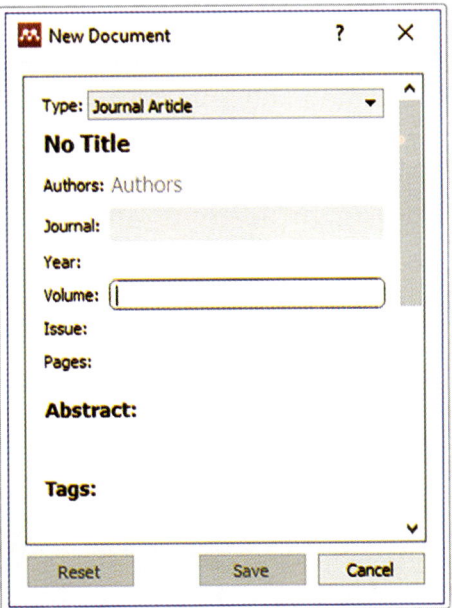

Fig 30. Mendeley desktop-Add entry manually Pop-up window

Source: Elsevier

3. There are various options in this window in the shape of a blank form. Select 'Type of reference/ document (Journal article, book, book section, case, computer programe, conference proceedings, encyclopedia article, film, generic, patent, report, thesis and webpages, etc.). After selecting the Type of reference, then all subfield will also be shown related to the specific Type of reference. After entering the bibliographic details, click on 'Save' button to include the reference in your Mendeley Library.

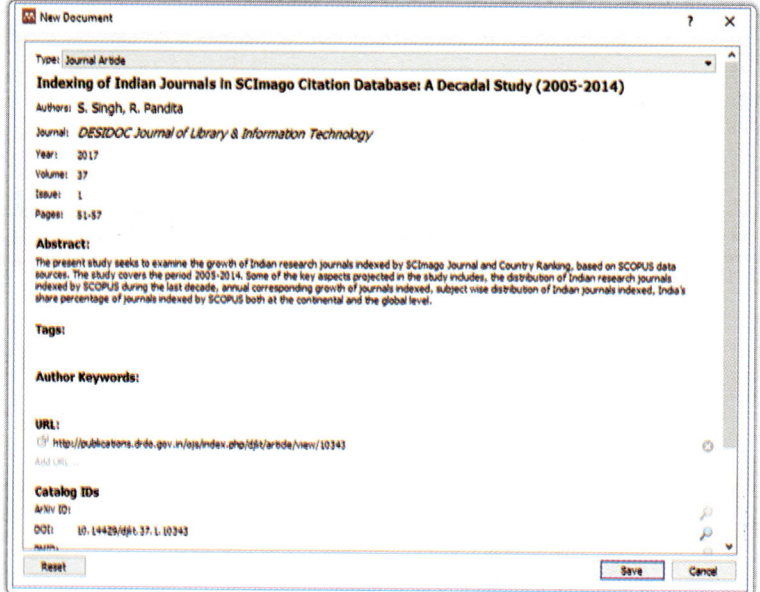

Fig 31. Mendeley desktop-Add entry manually Pop-up window

Source: Elsevier

4. The document's bibliographical details included in the Mendeley Library

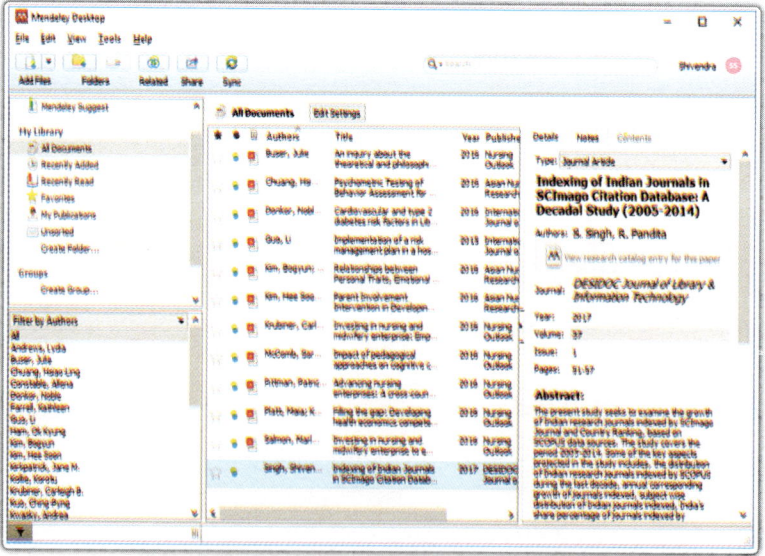

Fig 32. Mendeley desktop-Bibliographic details of added entry manually

Source: Elsevier

ADD REFERENCES BY IMPORT FROM BIBTEX, ENDNOTE, RIS AND ZOTERO FILES

To import from EndNote, need to have an EndNote.xml file which can be easily get from EndNote Export functionality and also from various online/offline database. You can add reference form .ris and BibTex file through Mendeley Import function.

1. Import from EndNote .XML file

    In EndNote Export the reference in .XML format

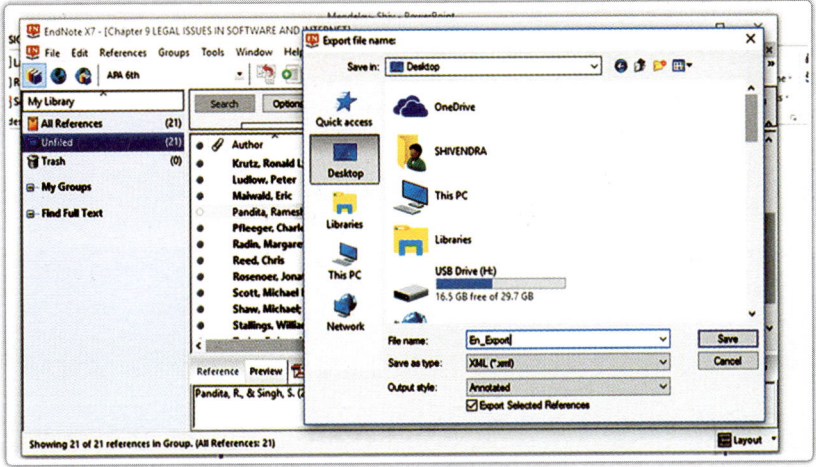

Fig 33. Mendeley desktop-Add reference entry from the other reference management files

Source: Elsevier

2. Go to Mendeley → File → Import → EndNote XML

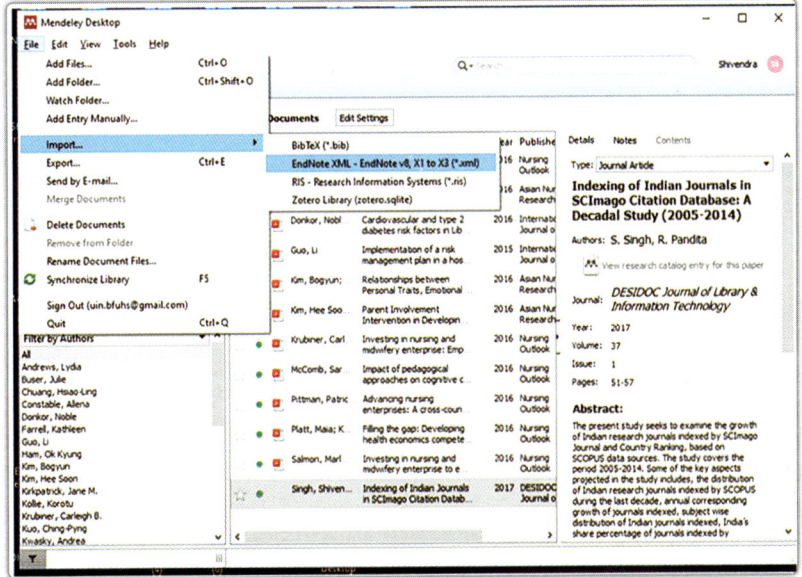

Fig 34. Mendeley desktop-Import menu for add reference entry from the other reference management files

Source: Elsevier

3. Go to Mendeley → File → Import → EndNote XML → Browse EndNote .XML file → Select → Open (Click on open button)

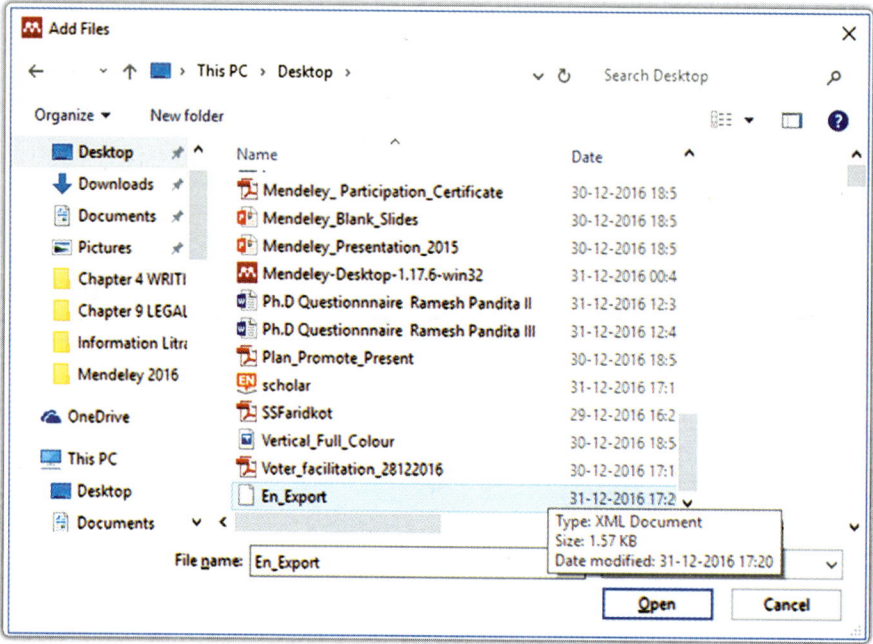

Fig 35. Mendeley desktop-Import menu for add reference entry from the other reference management files

Source: Elsevier

4. See the Imported File details in the Mendeley Library

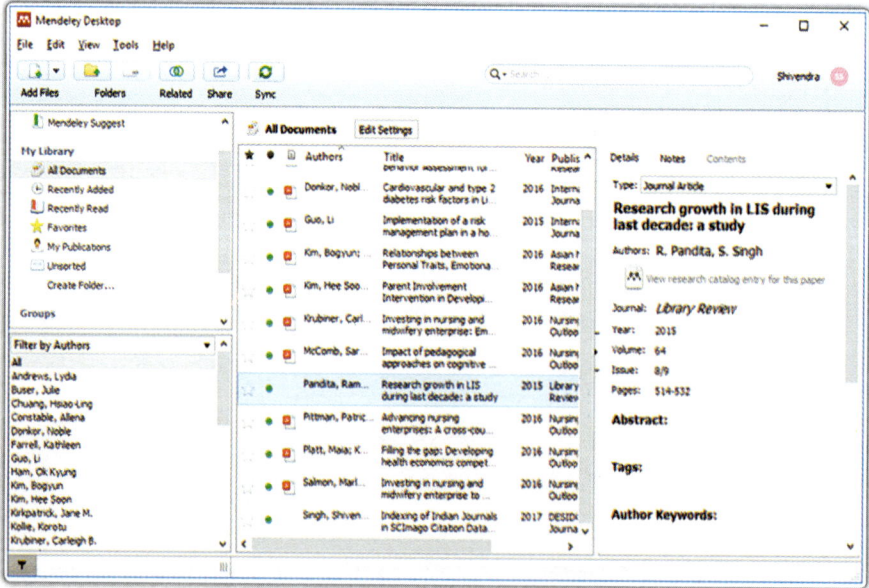

Fig 36. Mendeley desktop-Bibliographic details of imported endnote file

Source: Elsevier

BibTex File from Google Scholar

1. Go to any Internet Browser → open the www. scholar.google.com → Search the documents → Import into BibTeX → Save

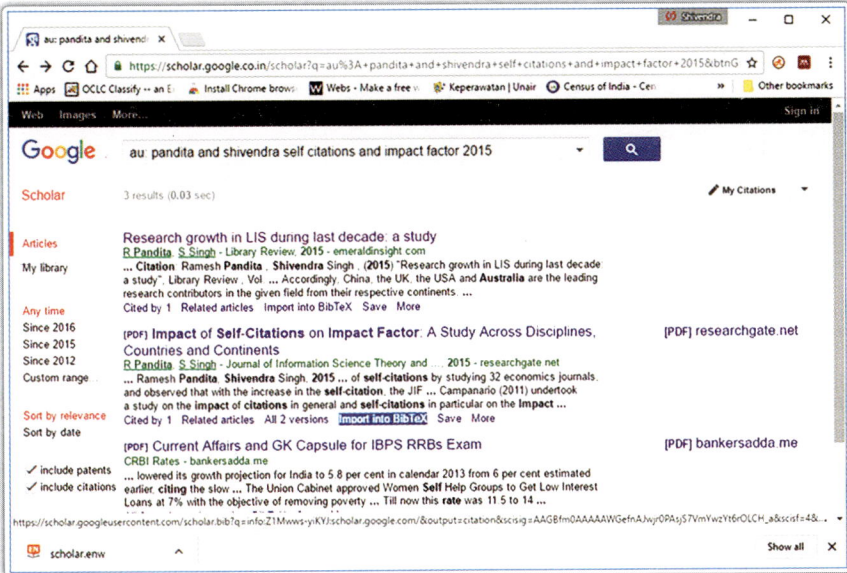

Fig 37. Mendeley desktop-Import from google scholar

2. Save the BibTeX file

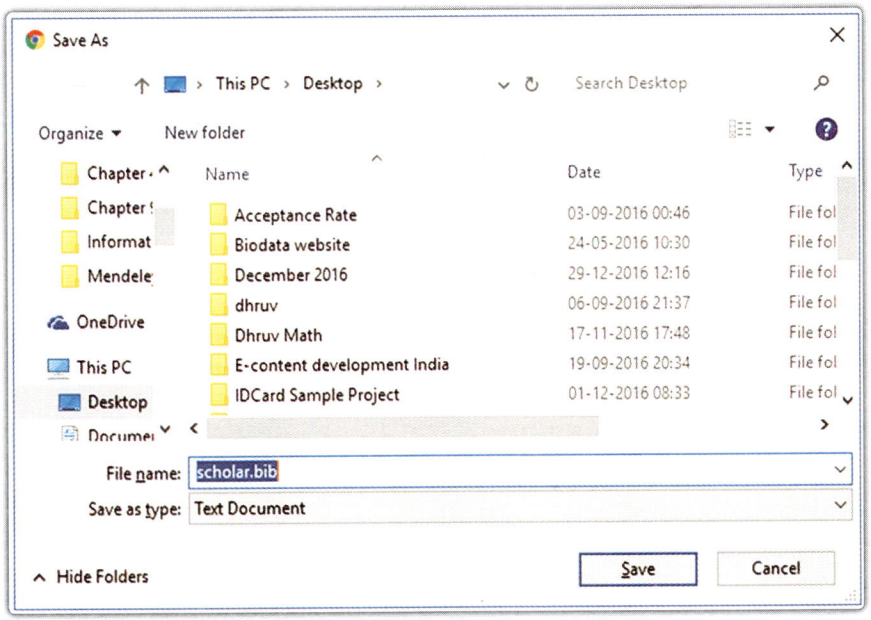

Fig 38. Mendeley desktop-Import from google scholar

3. Go to Mendeley → File → Import → BibTeX (*.bib)

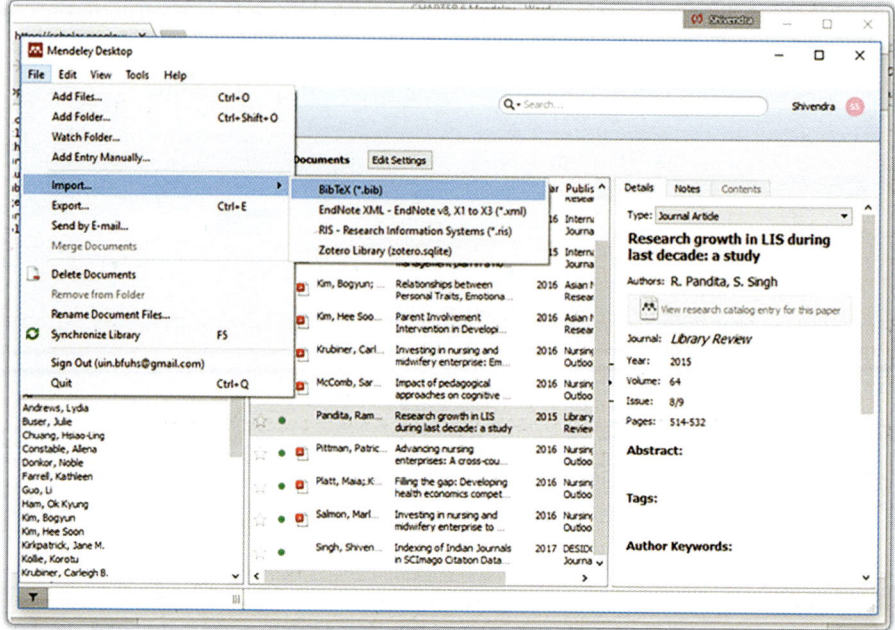

Fig 39. Mendeley desktop-Import from google scholar

Source: Elsevier

4. Go to Mendeley → File → Import → BibTeX (*.bib) → Browse bib file → Select → Open (Click on open button)

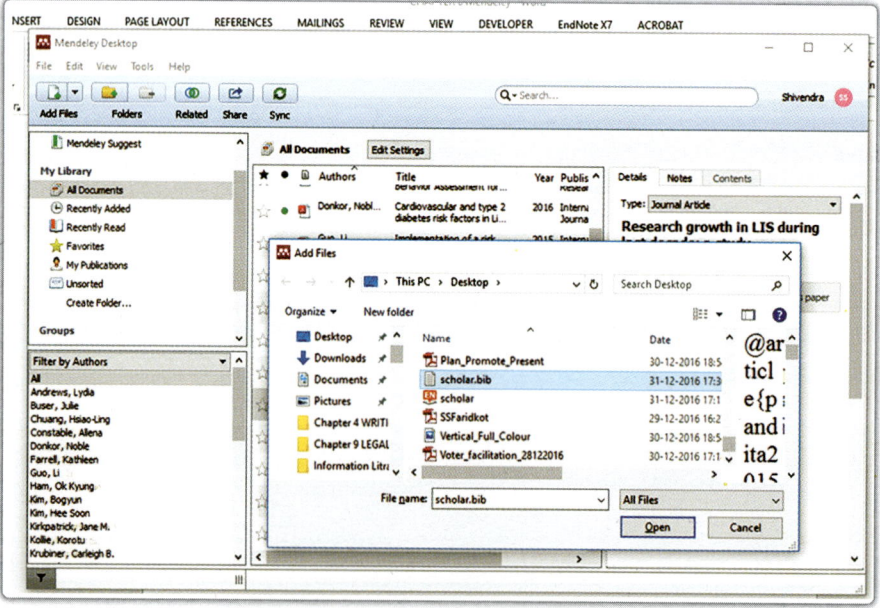

Fig 40. Mendeley desktop - Import from google scholar

Source: Elsevier

5. See the Imported .bib File details in the Mendeley Library

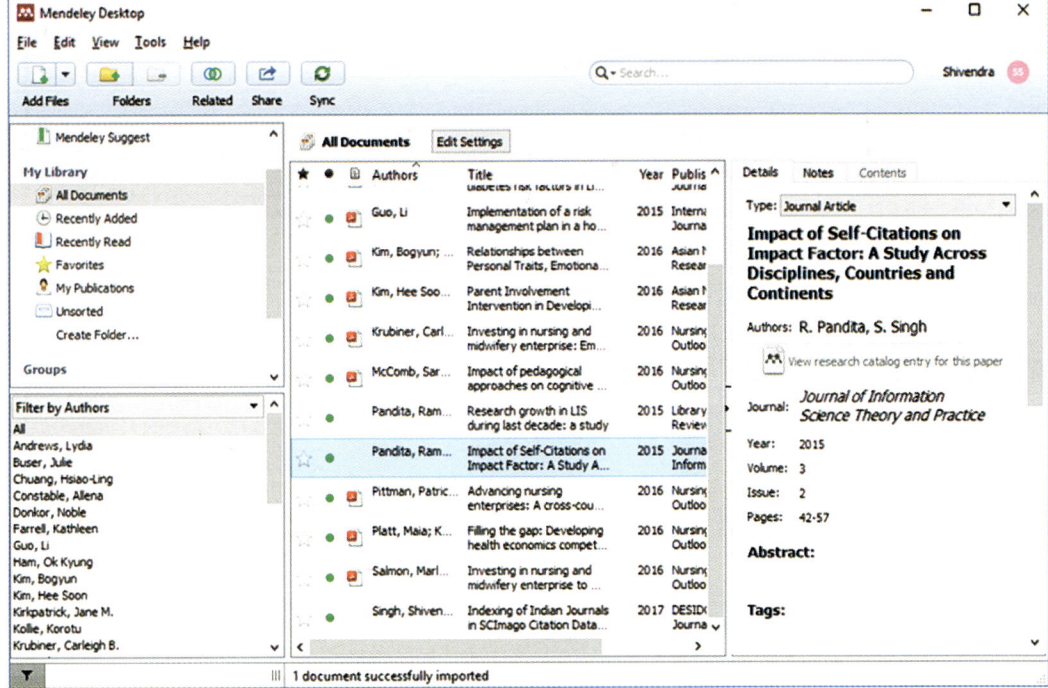

Fig 41. Mendeley desktop-Bibliographic details of imported file from google scholar

Source: Elsevier

6. The Same way you can add ris and Zotero files into the Mendeley Library through import functionality.

Add Reference Through Mendeley Web Importer

Import papers, web pages and other documents directly into your reference library from search engines and academic databases. Mendeley Web Importer is available for all major web browsers. (https://www.mendeley.com/reference-management/web-importer). The web importer help you when you are using any database like PubMed, Web of Sciences, and ScienceDirect or any other database. You have to use those database using the web importer features.

Install the browser extension from the Chrome Web Store

1. Go to Mendeley → Tools → Install Web Importer

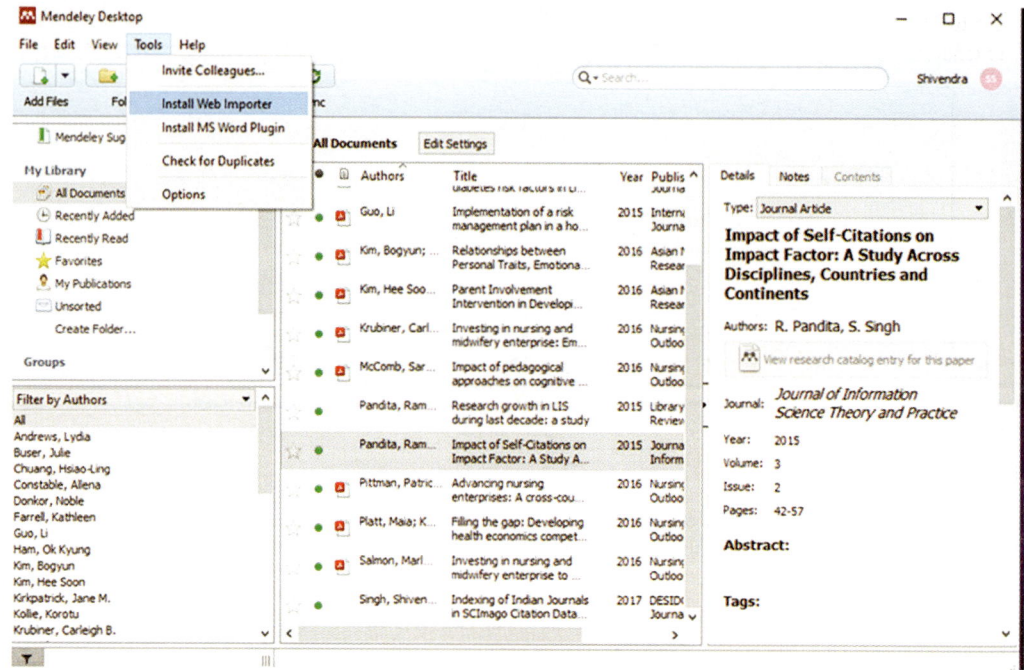

Fig 42. Mendeley desktop-Install web importer from the tool menu

Source: Elsevier

2. Automatically the Mendeley web importer page will be open on the default internet browser as one clicks on the "Install Web Importer" on the Tool menu.

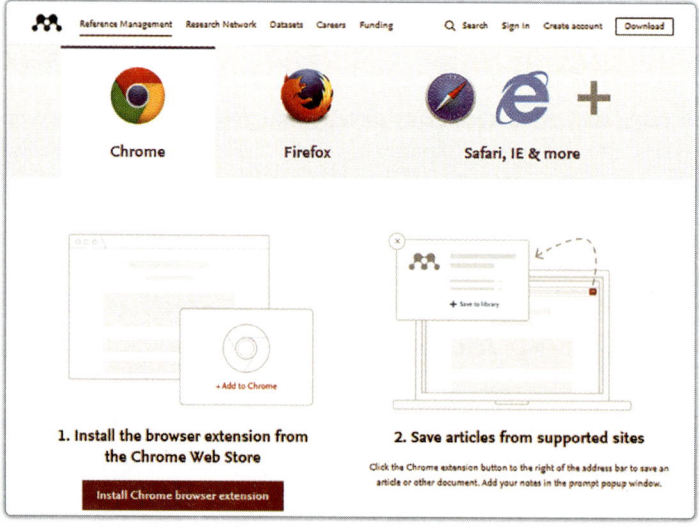

Fig 43. Mendeley desktop-Web importer for internet browsers

3. Go to Mendeley → Tools → Install Web Importer → Go to Web Browser → Install Chrome browser extension → + Add to Chrome

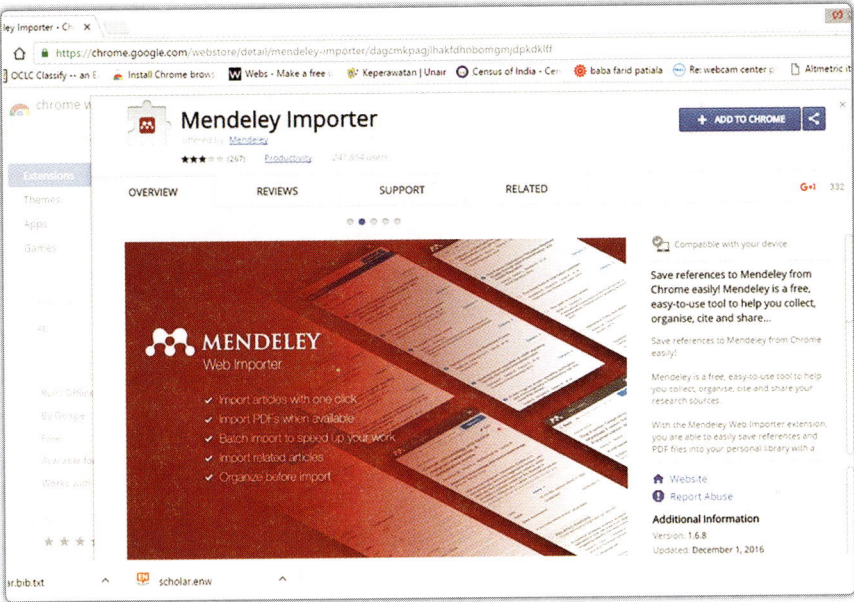

Fig 44. Mendeley desktop-Web importer for internet browsers

4. Asking to Add Mendeley Importer → Click on Add Extension

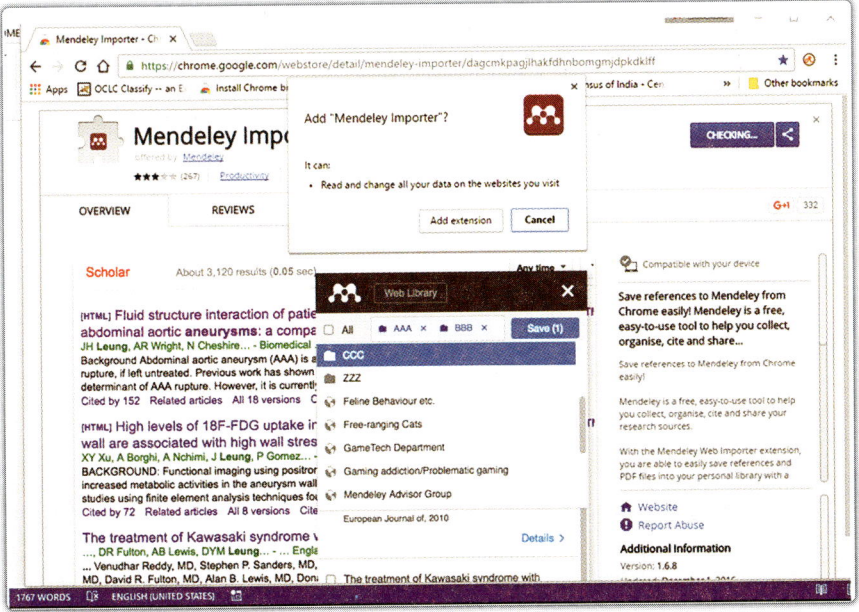

Fig 45. Mendeley desktop-Add web importer extension for internet browsers

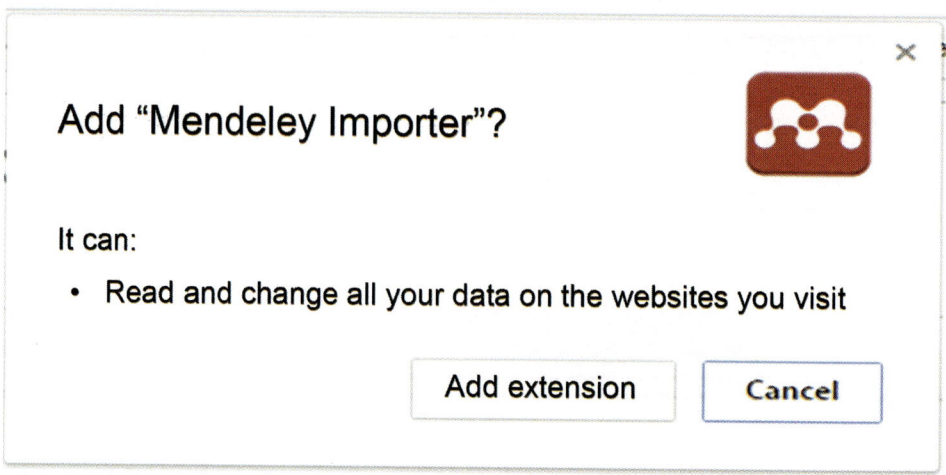

Fig 46. Mendeley desktop-Add web importer extension for internet browsers

Source: Elsevier

5. Mendeley Web Importer Added as extension

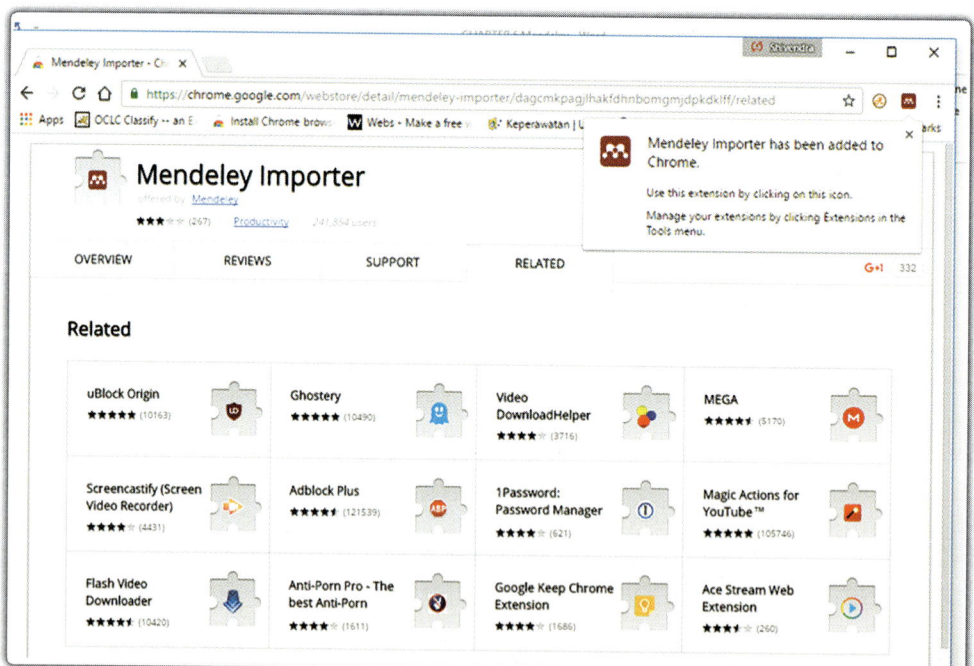

Fig 47. Mendeley desktop-Web importer extension at the internet browser

Mendeley Importer has been added to Chrome.

Use this extension by clicking on this icon.

Manage your extensions by clicking Extensions in the Tools menu.

Source: Elsevier

6. Mendeley Web Importer is added now in your default web browser. Now any document will be added in the Mendeley Library from the Internet Web Browser through the "Web Importer". Web Importer imports references and documents from an academic database with a single click on the extension button.

7. Add references from PubMed through "Web Importer"

8. Go to PubMed (www.pubmed.gov) → Search the document

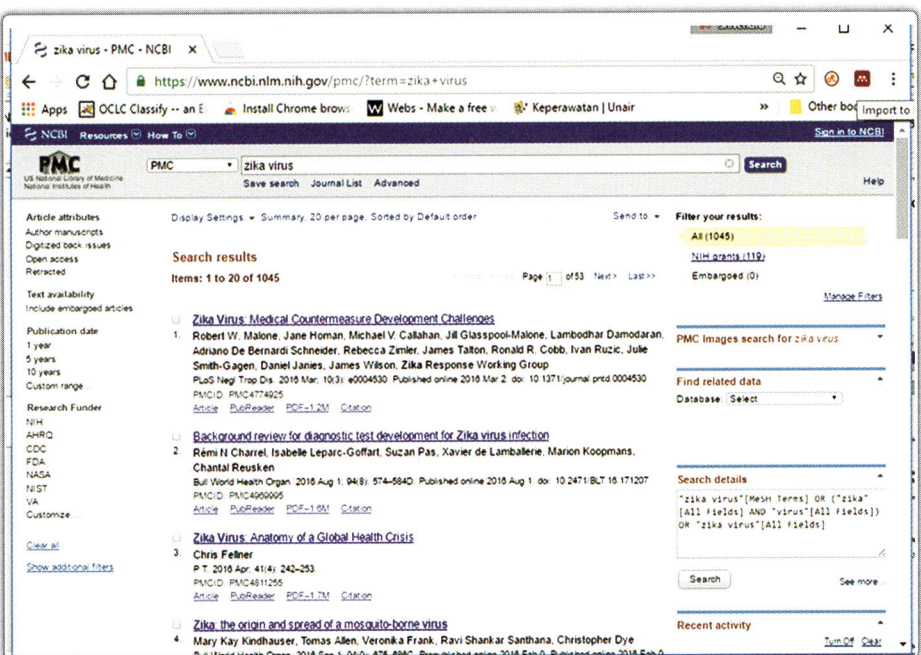

Fig 48. PubMed search window

9. Click on Mendeley Importer Extension on the browser to import these documents to Mendeley library

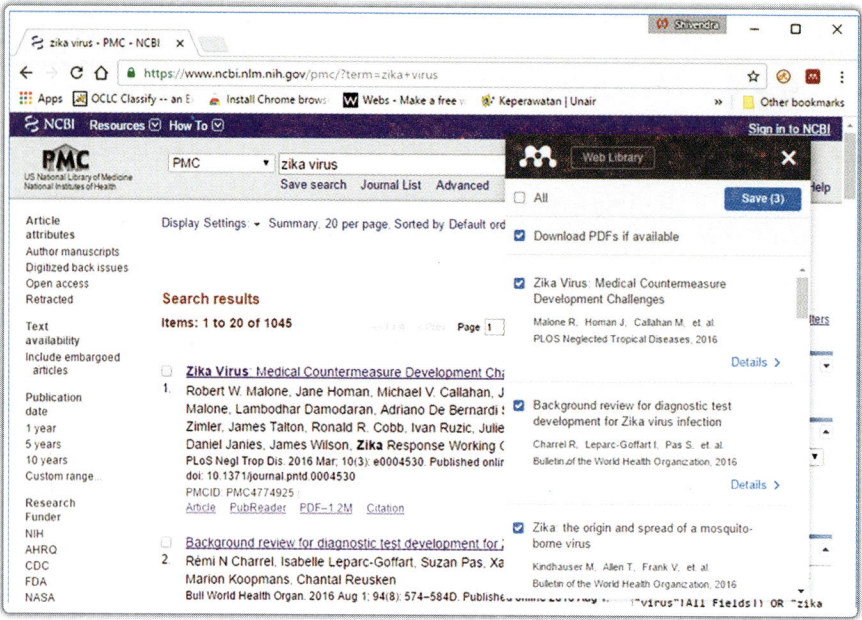

Fig 49. Save PubMed search result to mendeley through the web importer extension at the internet browser

10. Choose all the documents or selected ones and then click on save button. PDFs will be also downloaded, if available, when select one of the options given therein.

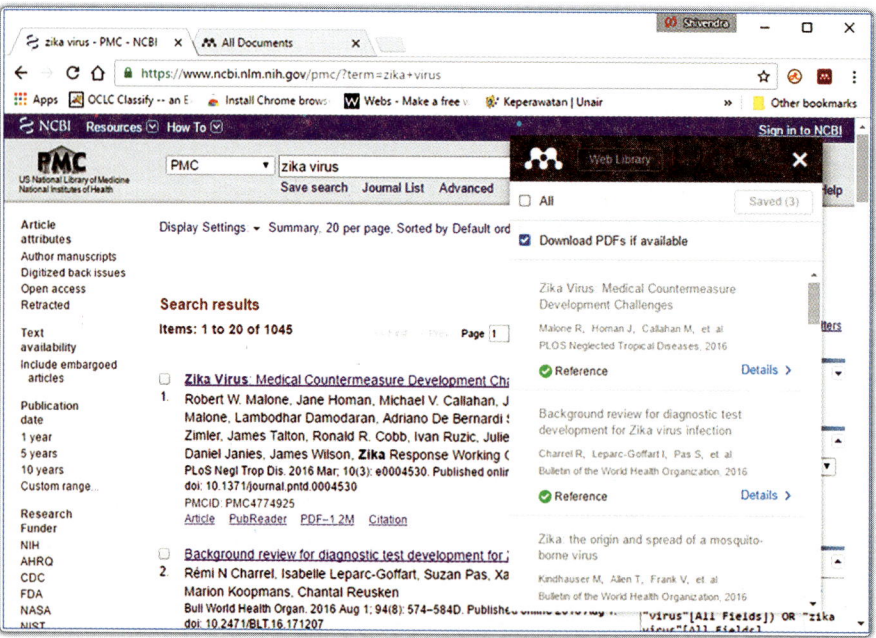

Fig 50. PubMed search result saved into to mendeley

11. All selected documents are stored in the Mendeley web library with PDF (if available).

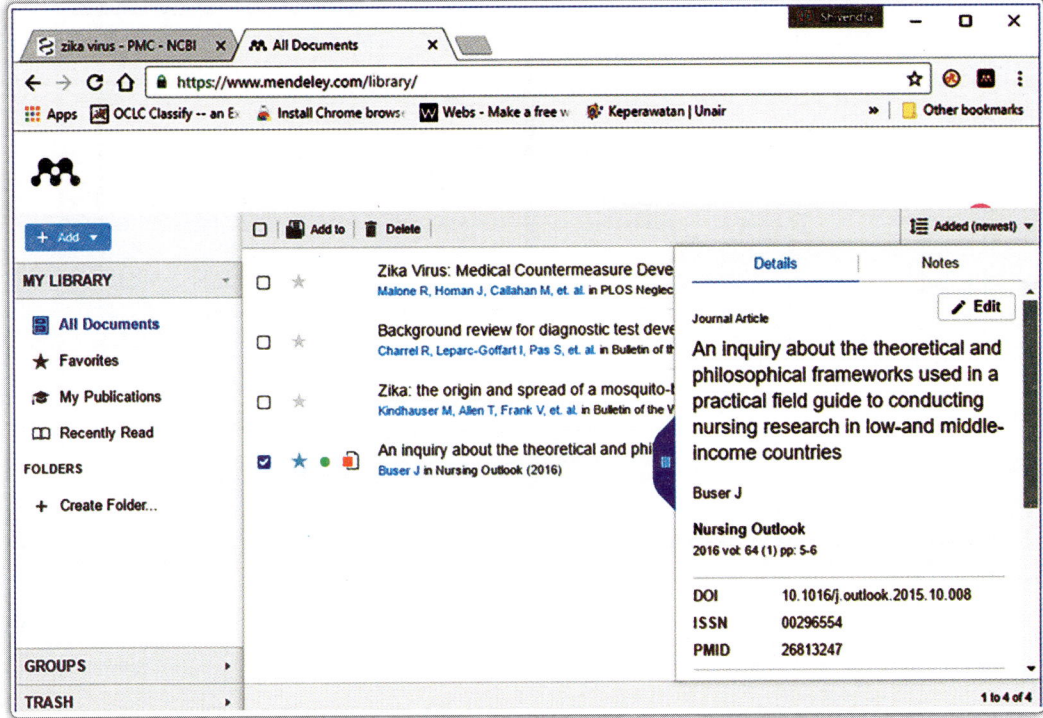

Fig 51. PubMed search result showing on the mendeley web

12. If you want to see these documents in your Desktop Mendeley, you need to be synchronized

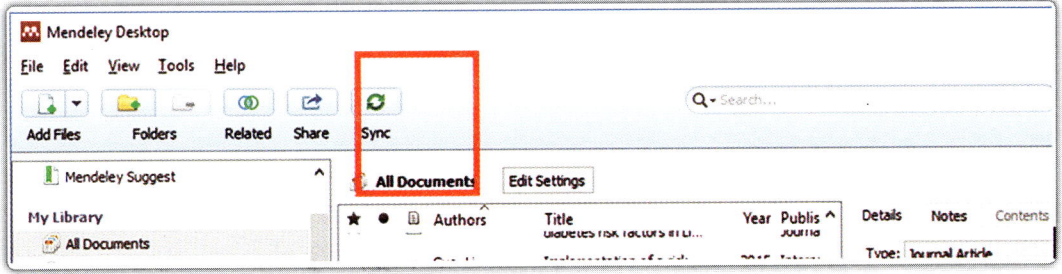

Fig 52. Mendeley desktop synchronized for web imported documents
Source: Elsevier

13. After synchronization

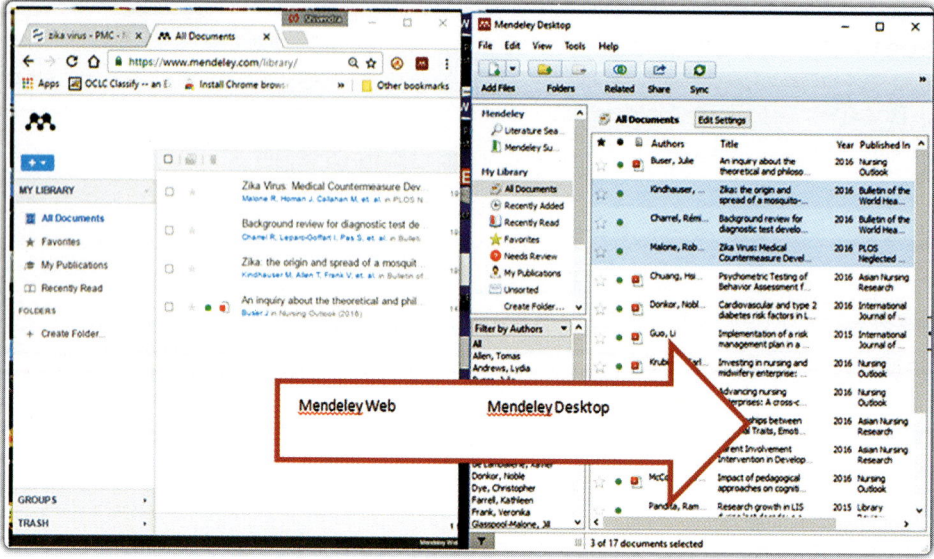

Fig 53. Mendeley web and mendeley desktop

Source: Elsevier

14. When you refresh the Mendeley Web in a web browser. All Desktop Mendeley Library references will be included in Mendeley Web library too (Synchronized).

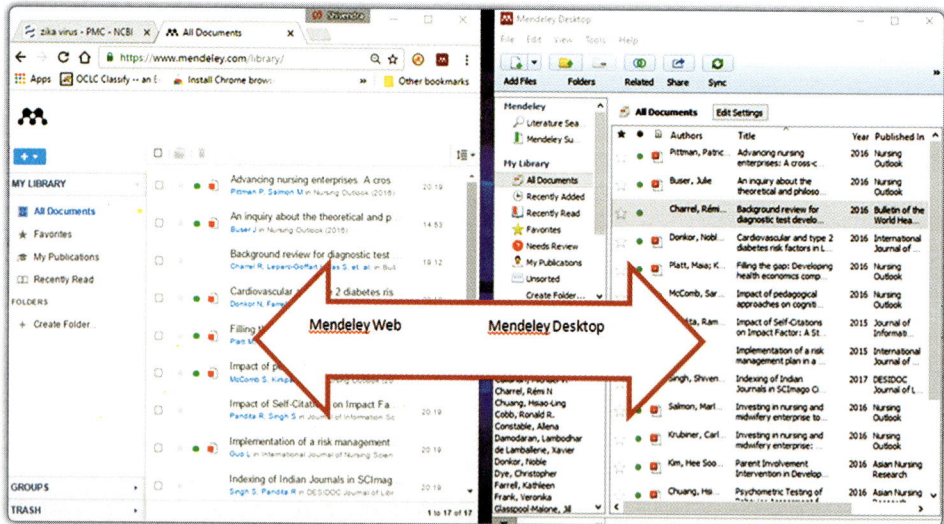

Fig 54. Mendeley web and mendeley desktop

Source: Elsevier

CITING REFERENCES

Install MS Word plugin

1. Go to Mendeley Desktop → Tool → Install MS Word plugin (MS Word and Outlook must be closed/ /not running for the install to this plugin. If you have not done before this, Mendeley will show the warning with this screen

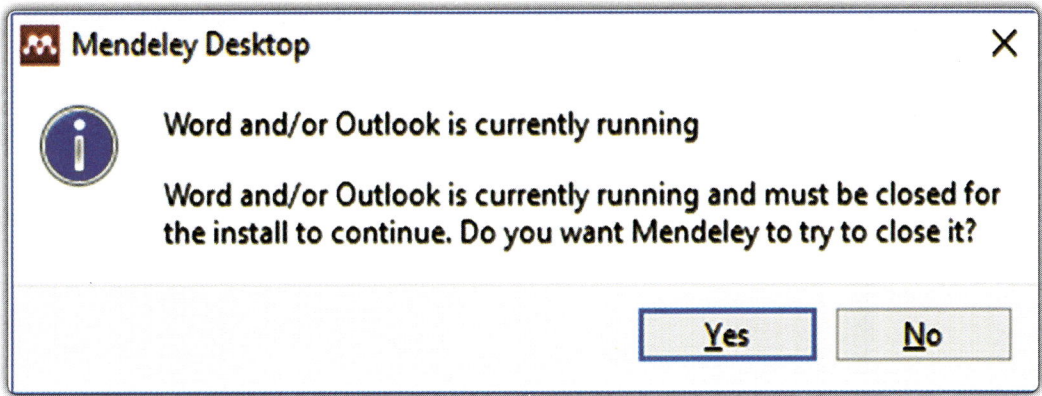

Fig 55. Mendeley desktop-Warning message window during the installation of ms word plugin

Source: Elsevier

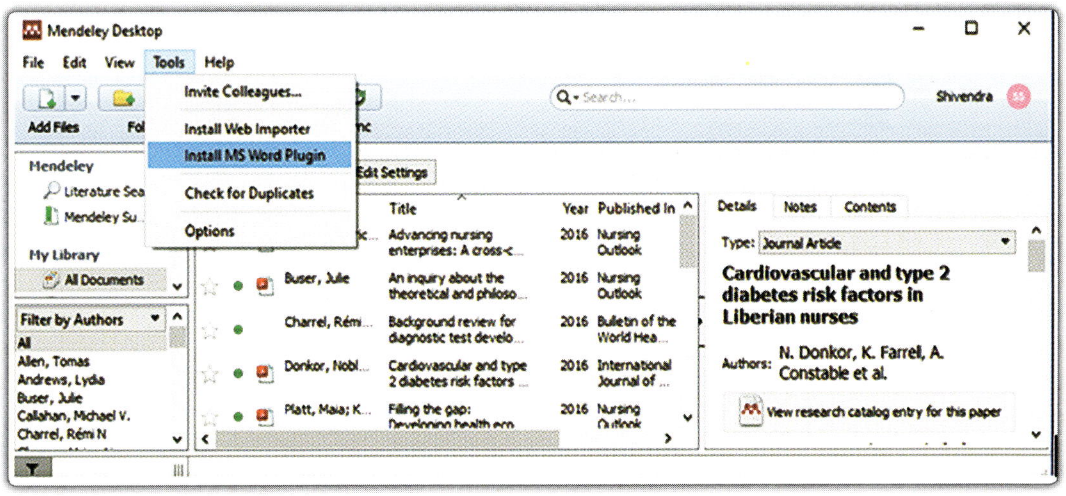

Fig 56. Mendeley desktop-Installation of MS Word plugin

Source: Elsevier

2. Plugin installation confirmation

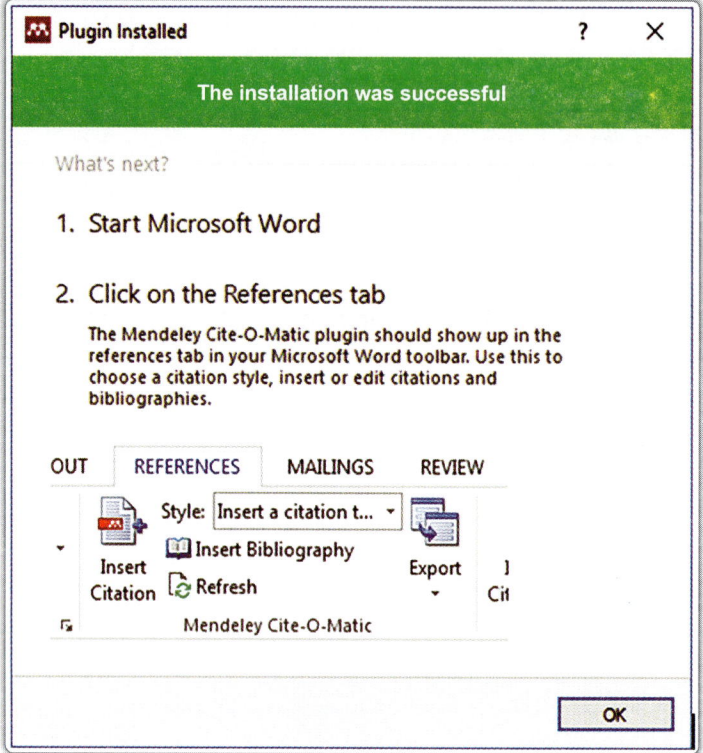

Fig 57. Mendeley desktop-Installation of MS Word plugin conformations

3. MS Word, when MS Word plugin is not installed

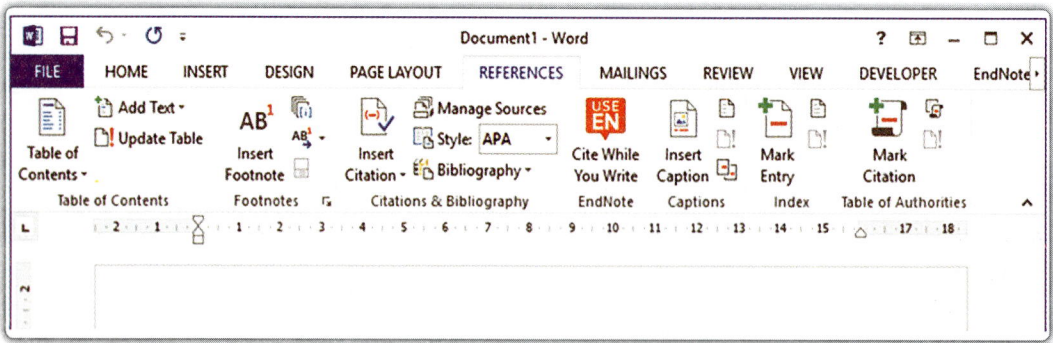

Fig 58. MS Word before mendeley MS Word Plugin installation

4. MS Word after installation of MS Word plugin

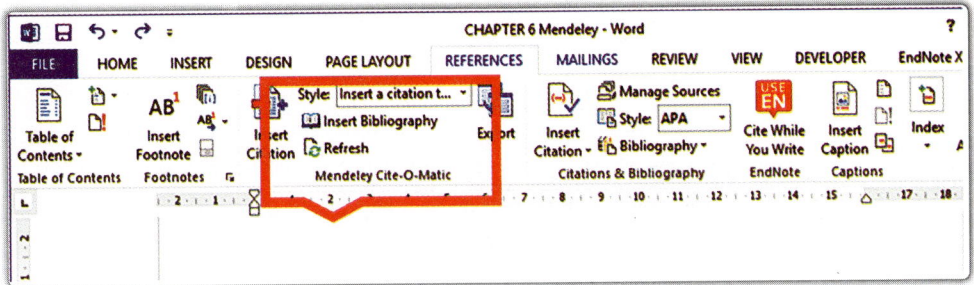

Fig 59. MS Word after Mendeley MS Word plugin installation

Added Mendeley Tool Bar (MS Word Plugin)

5. In MS Word opened file, put the cursor where you want to cite a document from your Mendeley Library.

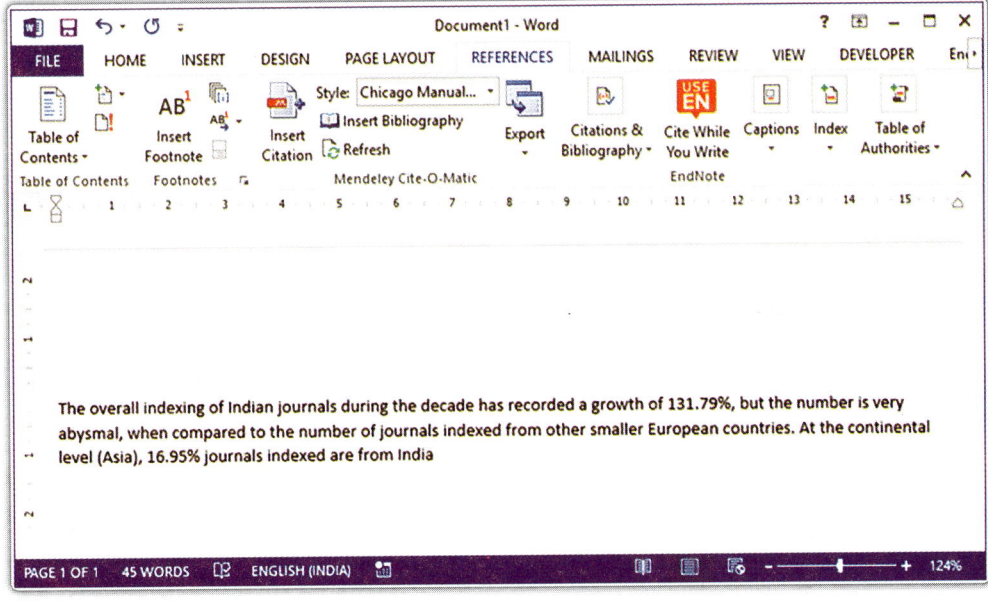

Fig 60. MS Word

6. Click insert citation in the Mendeley tool bar within MS Word

MS Word → References → Insert Citation (in Mendeley Tool Bar) → Go To Mendeley

Fig 61. MS Word-Insert citation from mendeley desktop

7. MS Word → References → Insert citation (into a Mendeley Tool Bar) → Go To Mendeley → Select the Document (in Mendeley Desktop Library) → Click on 'Cite' button to send the citation to MS Word file

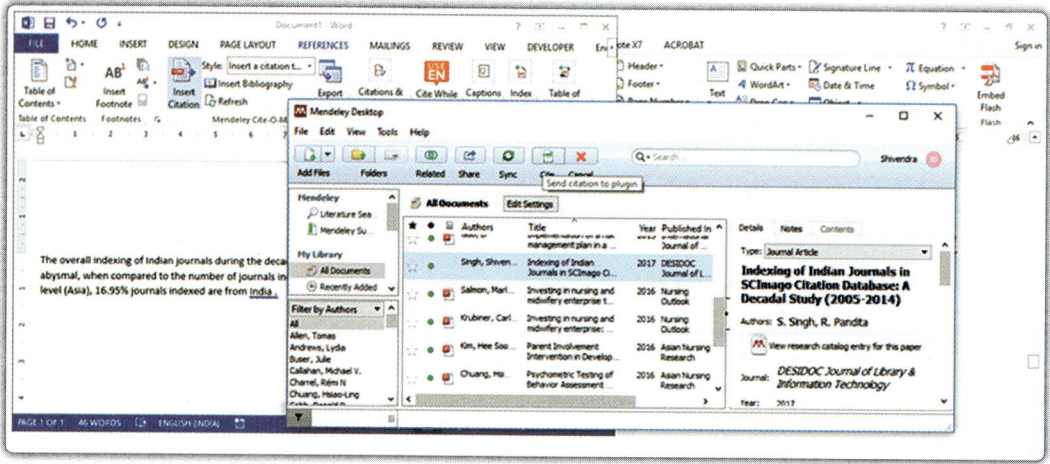

Fig 62. Mendeley desktop select the document

8. MS Word document after citing

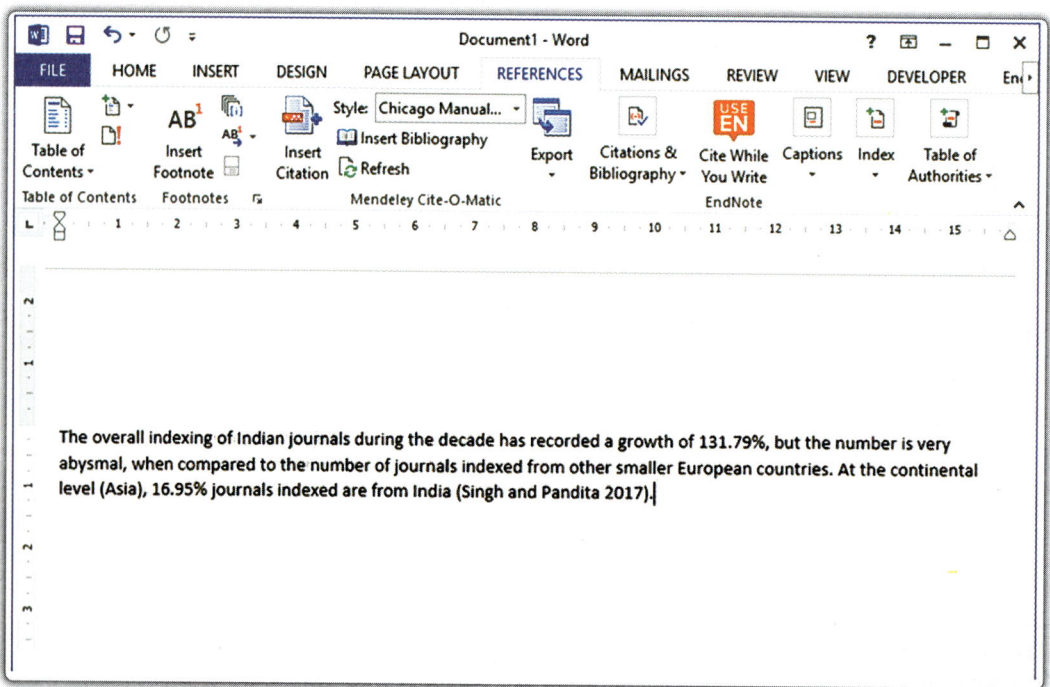

Fig 63. MS Word-Citation details of selected document from mendeley desktop

9. Insert Bibliography - After Insert Citation you can generate Bibliography

MS Word File → References → Mendeley Tool Bar → Insert bibliography

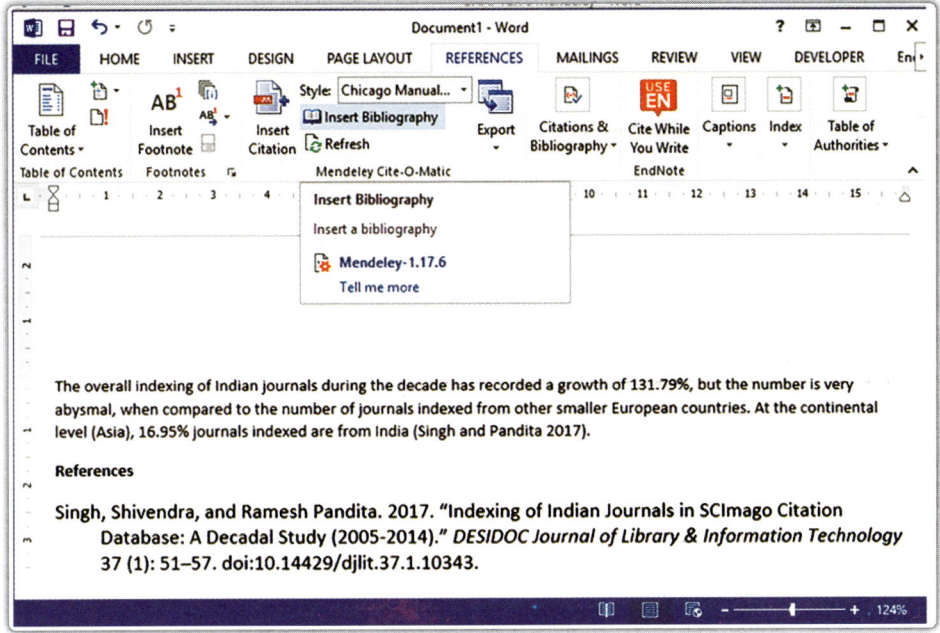

Fig 64. MS Word-Insert Bibliography or References

10. You can cite multiple documents. Just hold the Ctrl and left click the documents you want to cite. Various citation styles can be also selected from the Mendeley MS Word Tool.

There are many various features available in the Mendeley. Some of them are given below.

   i. Check duplicate documents
   ii. Documents can be marked read, unread and stared/un-stared
  iii. Search any documents with metadata
  iv. Tags
   v. Sharing documents and references
  vi. Create folder/rename
 vii. Create groups
viii. Invite members

CONCLUSION

Like EndNote, Mendeley is equally useful for reference management, used for citation and reference purposes. The software is equally popular among researcher for some added advantages to that of other proprietary and non-proprietary software. The premier version of Mendeley can be effectively used for referencing purpose and so it is useful for choosing the reference style. Mendeley helps to cite sources of information, from which they seek the information related to their study. The software helps in citing sources both within the text and under references and reduces the citing and referencing errors. It helps in matching the reference style of research journals, which authors are supposed to abide by.

BIBLIOGRAPHY

1. Download Mendeley Desktop. Manage and Share Research Papers. Mendeley [online] Available from www.mendeley.com/download-mendeley-desktop/ [Accessed December 15, 2017]

2. Elsevier Acquires Mendeley, an Innovative, Cloud-based Research Management and Social Collaboration Platform. [online] Available from www.elsevier.com/about/press-releases/corporate/elsevier-acquires-mendeley,-an-innovative,-cloud-based-research-management-and-social-collaboration-platform [Accessed December 15, 2017].

3. Elsevier buys Mendeley: your reaction. Higher Education Network. The Guardian. [online] Available from www.theguardian.com/higher-education-network/blog/2013/apr/10/ elsevier-buys-mendeley-academic-reaction [Accessed December 15, 2017].

4. Elsevier In Advanced Talks To Buy Mendeley For Around $100M To Beef Up In Social, Open Education Data. TechCrunch. [online] Available from https://techcrunch.com/2013/01/17/elsevier-mendeley-education/ [Accessed December 15, 2017].

5. Henning V, Reichelt J. Mendeley A Last.fm For Research? In 2008 IEEE Fourth International Conference on eScience. 2008. pp. 327-8.

6. Install instructions for Mendeley Desktop on Windows. Mendeley. (n.d.). [online] Available from www.mendeley.com/download-mendeley-desktop/windows/instructions/ [Accessed December 15, 2017].

7. Interview with Dr. Victor Henning, Mendeley - Scientific American Blog Network. (n.d.). [online] Available from https://blogs.scientificamerican.com/information-culture/interview-with-dr-victor-henning-mendeley/[Accessed December 17, 2017].

8. Mendeley. Free reference manager and academic social network. Elsevier. (n.d.). [online] Available from https://www.elsevier.com/solutions/mendeley [Accessed December 15, 2017].

9. Mendeley. How do I import my existing references from EndNote. (n.d.). [online] Available from http://support.mendeley.com/customer/en/portal/articles/227897-how-do-i-import-my-existing-references-from-endnote- [Accessed December 15, 2017].

10. Mendeley. Supported word processing software. [online] Available from https://www.mendeley.com/reference-management/citation-plugin [Accessed January 1, 2017].

11. Mendeley quick start guide. [online] Available from www.ucl.ac.uk/library/ [Accessed December 2017].

12. Paul Foeckler: Co-Founder of Mendeley. Doctorpreneurs. (n.d.). [online] Available from http://www.doctorpreneurs.com/paul-foeckler-interview/ [Accessed December 15, 2017].

13. Release Notes for Mendeley Desktop v1.8.1. Mendeley. (n.d.). [online] Available from from https://www.mendeley.com/release-notes/v1_8_1/ [Accessed December 15, 2017].

14. Victor Henning's brief guide to Mendeley. (n.d.). [online] Available from https://www.elsevier. com/connect/victor-hennings-brief-guide-to-mendeley [Accessed December 15, 2017].

15. Mendeley. Videos and Tutorials. [online] Available from https://www.mendeley.com/guides/videos [Accessed January 1, 2017].

Chapter 7

Legal Issues in Software and Internet

INTRODUCTION

It has been observed that people in general and researchers and academicians in particular pay a very little or less attention toward the legal issues involved with the use and purchase of software and other internet-based services. Accordingly, in the present chapter some basic discussions have been undertaken about the proprietary and non-proprietary software. The chapter deals with the software piracy, the legal issues involved with software use, software production and distribution and how actually even the so called informed sections of the society overlook these issues. Be they about licensing, copyright or various other legal provisions, which one is otherwise supposed to abide by and how actually the violation of same can easily bring disrepute to an institution, organization, agency or even an individual and if sued in the court of law by the proprietor, apart from paying damages, the offender may also face imprisonment. Concerns have been raised toward the growing menace of, cybercrime, cyber terrorism, computer vandalism, hacking, software infringement and many more.

Information technology has become an integral part of a modern day man and his dwelling methods. Information Technology has more or less created a dependency syndrome among people and in the process of its use, people have somewhere forgotten about the legal issues involved with the use of technology and perhaps how unintentionally people risk themselves by indulging into the illegal and unauthorized use of Information Technology and its allied aspects. In this chapter, we will talk about following few components, which involve legal issues, if not used under the legal cover.

- Software
- Internet
- Cybercrime
- Legal issues
- Software infringement
- Copyright laws
- Open source and free software
- The risks of using pirated software
- Conclusion

SOFTWARE

Before discussing about software, it is imperative to give an insight about computer or any other machine or an electronic gadget, which works on the pattern of computers. A computer as we know consists of two important and integral components what we call as computer hardware and software. Computer hardware constitutes the physical component of the computer or those parts what we can touch and feel. These include, the monitor, keyboard, CPU, etc. The software component on the other hand, is that part of the computer, which we cannot touch or feel, but it makes the machine to run or work. In the simplest terms, we can say, hardware component is the tangible part of the computer, while as the software part is the nontangible.

Software or computer software is a set of machine-readable instructions, which helps or directs a machine to perform a particular task. Computer software sends instructions to the central processing unit of the computer, which is also considered as the brain of the computer to perform certain tasks. Of late, we can see the market is flooded with different kinds of electronic gadgets, which work on the similar lines to that of computers. These computer like electronic gadgets constitute of both hardware and software components. Both the computer hardware and software cannot function in the absence of each other.

Types

Computer software are mainly divided into two types, one is system software, what is also known as an operating system and another is utilities software, what is also known as an application software. System software or operating system is mainly required to make the machine operational operational, in absence of which application software can not be run in the machine. Application software on the other hand, are those utility software, which are used to perform certain tasks on the desired lines. Each application software is designed to perform a specific function, and cannot be used interchangeably.

Enterprise software application is also application software, but is generally designed to perform and handle the specific operation and activities of that particular enterprise for which the software is generally developed. Software are developed by software engineers by pressing into service the high level programming languages. These software are written in machine languages generally known as machine code, which consists of binary values. In short, we can say each action a computer performs, is simply based on the set of instructions, passed on in the form of binary values or in machine code.

Forms

The softwares are mainly of Proprietary and non-proprietary form, the non-proprietary softwares are also know as open source softwares. Free software is one more form of the proprietary software, but is made freely available to the public. Non-proprietary software are also known as open source software.

- Proprietary software
- Non-Proprietary software/open source software
- Free software

Proprietary Software

Proprietary software are those software, which are commercially available in the market. These software are generally developed by software developing companies. Organizations involved with the commercial production of software do receive orders from various business houses to develop customized software keeping in view the in-house services and activities of these business establishments. Commercial software developers are also required to provide all sorts of services to all such organizations or individuals who purchase their developed software. Software developers also release patches, what we also call as updates from time to time to keep these commercially circulated software updated. Patches or updates are developed and released by software engineers generally when they find any flaw or scope for improvement in their developed software. It is the source code of these commercially developed software over which the software developers have copyright and this code is never released to public.

Non-Proprietary Software/Open Source Software

These non-proprietary software is also known as open source software. These software are not developed for commercial purposes, but they are widely used by user community all across the globe for their various routine activities. Open source software are released along with the source code. This source code is generally used by professional software developers for further improvement in the software efficiency and is being also used to customize the software as per the local needs of an institution, organization or an individual. Open source software are freely available and one can easily download them through the Internet and can press them into service. People need not to purchase or pay for these software in any form. These software generally do not involve the issues with regard to copyright, although it is there, but the software is released on the concept of copy left. Copy left concept in itself advocates about the freedom of improving the software as per ones local needs or for general use of the public.

Some Common Characteristics of Open Source Software are

- No royalty or any other fee imposed upon distribution
- Availability of the source code
- Right to create and modify derivative works
- No discrimination against persons or groups
- No discrimination against fields of endeavor
- All rights granted must flow through to and with redistributed versions

Free Software

There is a considerable difference between the open source software and the free software and both cannot be used interchangeably, which people generally do. The fundamental difference between the free software and the open source software is—the former is supplied free of cost to the users for use, but the source code of the software is not supplied to its users, while the later is also available free of cost, but the source code of the software is also made available to the user for further improvement and customization of software as per the local needs in the following ways.

- The freedom to run the program for any purpose
- The freedom to study how the program works and adopt it to your needs
- The freedom to redistribute copies
- The freedom to improve the program, and release your improvements to the public.

Software Piracy

Globally, there may be hardly any industry, making as much as lose on account of piracy as the software industry. Each individual user using any electronic gadget in his/her office, home, or in any other commercial establishment, which runs on soft technology, should be legal and should use a licensed product. Any individual, organization or business establishment using any software without purchasing it or without having a legal license is doing a criminal offence and can be challenged in the court of law by the proprietor. It is always desirable that properly licensed software to be used to avoid any legal issue. Each year loss worth billions of dollars is caused to proprietors on account of software piracy and the law enforcing agencies are finding it difficult to curb software piracy practices.

COMPUTER NETWORK

Network means a sort of connectivity between different things. We can have a human network, group network, community network, institutional network, academicians' network and many more. On the similar lines a computer network is a network in which a number of computers are connected together with the help of a network cable or via any other nonphysical medium like Wi-Fi, Li-Fi, etc.

Generally, we have three different types of networks in place, these are

- Lan: Local area network
- Man: Metro area network
- Wan: Wide area network

Local Area Network

Local area network is a typical network, which also helps one to develop a better idea about all the other existing computer networks. By local area network, we generally mean a network which can be established or created in a single building or even a single room by connecting different computers together with the help of hubs, cables (Ethernet) and adapters (Network). A simple and the smallest LAN can consist of two computers, which can be extended to any number of computers in a local area. This type of network is generally established to share resources within a particular organization or office, but not with the rest of the world. It is actually the server, which is being accessed by the rest of the clients.

Metro Area Network

Metro area network is a network, which may run across a whole town or across the entire city. A cable TV network running across the entire city can be considered as a MAN. It is the large network when compared to local area network. This kind of network runs on similar lines to that of a LAN. A metro area network can also be formed or established by connecting various local area networks together. A network running across district headquarters or any other larger area like a region or whole state is a wide area network.

WIDE AREA NETWORK

Wide area network is the largest of all the networks, which can run through entire country, region and continent. A wide area network is formed by interconnecting various LAN's and WAN's together. Internet or www is considered as WAN.

Of late, IT professionals have added a few more specialized networks to the list, these include

- Virtual private network (VPN)
- Personal area network (PAN)
- Storage area network (SAN)
- Enterprise private network (EPN)

INTERNET

Once you know about the computer network, it is easy to understand the internet. Internet is a network of networks spread across the length and breadth of the globe. This global network is spread across over 196 countries of the world, connecting millions of computers with one common thread to exchange information, data and other relevant information among any number of users. As per the data available more than 3.73

billion users across the world use internet and having access to over a billion of existing websites. There are different types and kinds of market players working in the communication industry and are actively involved with providing internet services to users in a variety of ways with internet protocol services as the most preferred way to access internet.

- Internet, purely works in the decentralized form, there is no single individual or organization which owns it. As said above, that data remains actually scattered and stored across the different web servers and moment all these servers are connected with each other via the internet, the data or information becomes accessible to the information seeker.

- Many a times, it is being seen that people use internet and www interchangeably, while as the fact remains these are two different things. Internet as said, simply means the network of computer across the globe, while as the world wide web or web is an information sharing model to access the remotely located information on the internet.

CYBERCRIME

Crime has always moved parallel to the societies in every era. Crime was prevalent in society, even during the period when men used to live as nomads, hunters and food gatherers, thereon, when the men started living in settled colonies, the crime move alongside. Now, since the society has moved from physical to virtual world by embracing information technology, so has crime added one more face to its forms? Cybercrime has become the buzz word in the IT world. Experts working in the IT sector are even of the opinion that cybercrime has much more far reaching and adverse consequences than the physical form of crime. The experts are of the view that there is far greater need to regulate IT world by putting in place all sorts of legal and ethical provisions whereby crime be prevented from spreading its tentacles unabated.

Cyber criminals, unlike the ordinary criminals are highly talented, having mastery of software technology. It is actually the ethical consideration, which regulates the professional behavior of each individual, on the other hand is the same ethics which are being actually violated by these cyber criminals. These criminals instead of using their talent for creating and welfare work of one and all get somewhere deviated and indulge in undesirable acts. Cybercrime is committed by using a computer and the internet to steal data or information by making use of malicious programs to make illegal imports, etc.

Categorization of Cybercrime

Unlike the physical crime, cybercrime remains confined to the computers, it is the computer which is being targeted as victims, and again it is the computer, which is being used as a weapon.

- The computer as a target
- The computer as a weapon

Types

There is not a single or any defined form of cybercrime, perhaps it will not be inappropriate to say that there are as many forms of cybercrime, as many there are ways we use computers in our routine life. Every new day, we come up with a new software application form, whereby we are being enabled to undertake any particular task in the soft form, which otherwise we used to undertake in its physical form. So thereby, each individual software application is a potential victim of the cyber criminals. Some of the common types of cybercrimes prevalent in the society, which the internet user is quite familiar include:

- Hacking
- Denial of service attack

- Virus dissemination
- Computer vandalism
- Cyber terrorism
- Software piracy

HACKING

- Computer hacking is directly related to the computer security if the system is secure from all sides, preferably by having all security features active, be it in the form of firewalls, anti-viruses, password protection, access control system and other application security aspects, the computer system is less vulnerable to external threats or becoming a victim of cyber criminals. If a computer operator is not taking every care of the computer security aspects, his/her computer is more vulnerable to hacking.

- A computer hacker on the other hand is an individual, who is motivated toward unethical practices and tries to exploit the weakness of other computer systems, which he is able to access on a computer network. There is no single defined motive behind a computer hacker, it can be for personal reasons, it can be for professional reasons, it can be simply a revenge, it can be money, challenge and enjoyment, etc. are the few other common reasons seen in this direction.

- Arguments are also being made about the ethical and unethical hacking. Ethical hacking also refers to those practices whereby some vital information, which may be of social or the hacking undertaken for the lager interest of the society, is termed as ethical.

- Ethical hackers are more often referred as white hats, as they are generally able to crack the codes and retrieve desired information from the antinational or antisocial elements under the legal cover, where as unethical hackers on the other hand are known as black hats.

- Illegal intrusion into a computer system and/or network.

Computer Vandalism

Computer vandalism is more associated with professional or personal rivalry. In this type of cybercrime, a criminal hacks the computer of targets or victims with the intent to damage the data or other vital information, including steeling, etc. Cyber criminals associated with such criminal practices generally damage and hurt the professional and other interests of the target with no real personal gain, except fostering animosity. Computer vandalism involves the following practices.

- Damaging or destroying data rather than stealing
- Transmitting virus.

Cyber Terrorism

There is a lot of ambiguity over the use of term cyber terrorism. IT professionals and experts term the cyber terrorism as the cyber warfare and cannot be correlated with the conventional or physical forms of terrorism, as no physical harm can be done to people by using a computer. Although, there are growing instances of the internet being used as one of the easiest means of information communication by the terrorists across the world, mostly by using cryptic languages, hence it is termed as cyber terrorism. Computers can be easily used as soft easy mediums to terrorize the people on various grounds, so in a way the term can be justified to a more or lesser extent. Still more, others define it simply as a practice whereby networks or the internet is used for intentional destruction or damage of on ideological grounds. The growing concerns shown about the cyber terrorism are mostly because, the motives are to harm the interests of the society using IT as as a tool. Eugene Kaspersky, the founder of Kaspersky Lab is of the view that cyber terrorism is the more appropriate term to be used for cyber warfare, for the fact that one remains clueless in this warfare and the possible future attacks.

The internet is a safe, favorite choice for terrorist activities due to the free access, lack of supervision, anonymity and unlimited information dissemination. (Kolbai Musave, deputy chief KRYGYZ Republic). Besides, there is no second thought in it that any form of practice, which targets and terrorize people on one or the other pretext is a terrorism of its own kind.

- Use of Internet based attacks in terrorist activities
- Technology savvy terrorists are using 512-bit encryption, which is almost impossible to decrypt.

SECURITY TIPS

It is not that a computer connected to the internet is always vulnerable to cybercrime or cyber. It is truly said, prevention is always better than cure. There are various preventive measures in place, which can help a computer user with a great deal in dealing with all such cyber threats. It is always advisable to a computer professional or anybody using computers for different purposes to take the following measures to avoid any cybercrime incident:

- **Use antivirus software:** It is always advisable to use a good quality or tried and tested, antivirus software as a preventive tool
- **Password change:** Professionals using computers regularly for both the official and personal purposes, should develop the habit of changing their passwords frequently
- **Insert firewalls:** Working in network environments always warrants having a firewall in place to regulate the traffic of incoming and outgoing network traffic and also ensures a secure and tested network access
- **Uninstall unnecessary software:** Undesirable software installed in a system, which is sensitive for information handling and data storage should be avoided. These software more often are vulnerable to hacking and soft targets for computer virus. By uninstalling undesirable software, your system may respond to your commands more positively and more speedily
- **Maintain backup:** It is always advisable to take regular backups of the data, as the loss of same can easily turn riches into rages. So be careful and adopt this preventive practice
- **Check security settings:** Security setting of a machine should always be ensured, it would be appropriate to seek the help of your system manager to ensure the security setting of your desktop
- **Stay anonymous: Choose a genderless screen name:** Anonymity in itself acts as a level of security, so there is no harm if it can help a professional to add one more security level to his/her profile
- **Never give your full name or address to strangers:** Internet is laced with both trusted and deceitful websites, as such, as long as one is not sure about the authenticity and the genuineness of the website, one should never disclose his/her true credentials. Even it is not advisable to register with every website here and there, as this data can be equally misused by the both the cyber and professional criminals
- **Learn more about Internet privacy:** Privacy is the right of each individual and nobody is supposed to unnecessarily stroll in the domain of others without their prior permission. We maintain privacy in different areas in our routine life and so is required to be abided while using the internet and its other applications

LEGAL ISSUES

Legal issues are involved with almost each and every kind of activity we undertake, therefore one has to be very careful every time about the legalities involved with the use, abuse and misuse of information technology. Most of the time it is being seen that IT professionals in the under developing and other third world countries are generally using pirated software. Software piracy has become one of the greatest concerns in the IT world. Even the so called developed countries are not elusive of software piracy. Every country in the world has law to deal with ethical use of IT and its related activities. Any misuse or abuse of IT in any form is challengeable in the court of law and the person found guilty of violating any such IT related laws can face severe punishment, including imprisonment.

Most of the legal issues involved with the use of IT and its byproducts are related to

Copyright

- Copyright is a legal right given to an individual under legal cover to use, distribute his creative work or any other product, which is the outcome of one's intellectual application. Copyright and intellectual property right is always seen together, for the fact that copyright is actually granted for the intellectual property. Accordingly, in the field of IT there are millions of such byproducts, which are being used by people in a variety of ways and the copyright of all such products rests with those who invented them.

Copyright Laws

- The Berne Convention for the Protection of Literary and Artistic Works, usually known as the Berne Convention, is an international agreement governing copyright, which was first accepted in Berne, Switzerland, in 1886.
- The Buenos Aires Convention is a copyright treaty signed at Buenos Aires, Argentina, on August 11, 1910, providing mutual recognition of copyrights
- Copyright laws on the illegal use of software are covered under Section 117 of the Copyright Act of the USA.
- End User Licence Agreement (EULA)
- The No Electronic Theft (NET) Act was signed into law in 1997
- Confers copyright protection for computer content and imposes sanctions for infringement
- The 1998 Digital Millennium Copyright Act (DMCA) contains several provisions
- Protects ISPs from acts of user infringement
- Criminalizes the circumvention of software protections.

Categories of Copyright Laws

- Civil law
- Criminal law

Owner's right of copyrighted work

Following are the rights of the owner and copyrighted work

- To reproduce the work
- To distribute the work
- To create derivative works
- To perform the work, or
- To display the work.

Copy-left

The concept of copy-left is not that widely popular among general masses, but the concept seeks its origin in the copyright itself. The concept of copy-left started receiving more recognition with the coming into being of the concept, open source. Open source software are generally provided with the source code by its developers so that others may improve upon it. Thus, in a way these developers leave a copy of their right with every other individual whosoever wants to further work on any such product? So this way each researcher after improving the product, leaves a copy for further improvement of the products, hence the concept copy left gained the acceptance in the society in general and researchers in particular.

Patent

Patent unlike copyright is granted in the field of inventions only, wherein a product invented by an individual cannot be made or reproduced in any form by any other individual without the prior permission of the patent holder. Inventors having patents in different field make commercial gains by licensing them to third parties for commercial purposes.

Trademark

Trademark is a physical, recognizable sign, which authenticates the product of a particular brand of a particular company. Trademark is being widely used by each kind of manufacturing industry or even on the consumer products to make all such products easily identifiable with the intent to promote the brand. Corporate houses use their logos or trademarks for advertisement purposes by displaying them on huge hoardings across cites, etc.

Service Mark

Trademark and service marks are mostly mixed-up together as one and the same thing, but the fact is that these two different marks recognize two different things. Service mark is more popular in the US and the other developed countries, whereby different kind of service provider use service mark as an advertisement to promote the type of services they are providing to the public. Of late, we can see that service marks are being used in a great deal in the Indian subcontinent as well. In fact, it won't be inappropriate to say that the service sector is emerging as a huge market in the each kind of economy.

Sound Mark

Sound marks are also used as part of the trademark to identify a product or service of the particular industry. The sound marks are generally used by the organizations involved with the entertainment industry. It is a very common sight to see any motion picture beginning with its sound mark to establish a sort of legacy that this particular motion picture is from a particular banner.

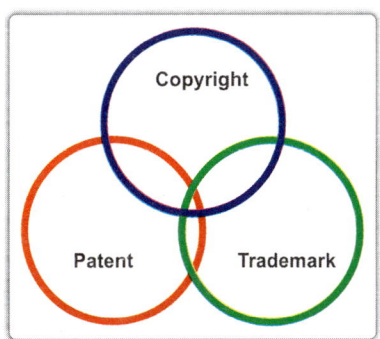

DIGITAL PROPERTY

The law protects intangible or intellectual property through three basic mechanisms:

1. Patent laws
2. Copyright laws
3. Trademark laws

SOFTWARE INFRINGEMENT

Infringement in the simplest terms means a violation of laws, rights, treaties, MoUs, etc. When we talk about the software infringement, it simply means the violation of the copyright laws by making use of the works of others, duly protected under law without the permission of the copyright holder. Infringement also amounts to the crime done intentionally, despite having the complete knowledge about the use or misuse of any product or service will involve copyright violation.

There are global concerns over the copyright infringement, as there are innumerable ways wherein infringement can occur. Some of the common practices associated with software infringement include.

- Counterfeiting occurs when illegally copied software is duplicated and distributed on a large scale
- Countries with weak software copyright enforcement cost software owners billions of dollars in revenue loss
- Globally, over 60% of the software sold is an infringing version.

Piracy Concerns

Piracy is not limited to the one or two fields. In different fields the term may be used differently. In the simplest terms, piracy means theft. Software industry as on date may be the largest industry in the world, which may be losing money worth billions of dollars and still there appears no stopping of this piracy market. People illegally use, copy and distribute the software among the wider population without having any ownership rights. Although, the majority of the software in the market comes as licensed products, but most of these products get easily decoded by the IT professionals for its wider use, ultimately resulting into the economic loss to the proprietor.

Apart from enforcing laws and regulations, there is an equal need to sensitize the people about the piracy concerns and the penalties, which the violators of law may pay for indulging in such acts. People should be made aware about the

- Laws regarding software piracy
- Penalties for pirating software
- Responsibilities as a software user
- Economic impact of software piracy
- Preventive measures.

Risks of using Pirated Software

By and large people appear least bothered about using the pirated software, even there are computer users in the world who pay for the original software products, but are supplied with the pirated ones and they seem to be less interested in ascertaining the authenticity of the product. Computer users must know that using pirated software may involve a number of issues, which may have far reaching consequences. Some of the common risks involved with the use of pirated software, which users must be aware of, are

- It is illegal to use a pirated software
- Low on features
- Increased cost/low productivity
- Malware attacks
- Vulnerability
- Information disclosure and data theft
- Extortion using ransomware
- Unsecured business environments
- Network effect and reputation risks.

SOME FACTS

- Everyone benefits from a healthy computer software industry
- When a few people steal software, everyone looses

SOME LITIGATION CASES....

- U.S army settled a lawsuit with Texas based company Apptricity in 2013
- Few days back Ex-VC of Delhi University sued under plagiarism

DIFFERENT LAWS

Each country has its own laws and bylaws while dealing with the legal issues involved with each individual kind of act. Accordingly, each individual country has put in place all such digital laws which have been formulated to regulate the ethical use of IT, internet, software and all other digital technologies.

When we talk about the digital laws, it means we are dealing with the ethical and unethical practices involved with the digital use of information technology. Some of the common ethical issues involved with the digital world pertain to the copyright, piracy, software infringement, etc.

Ethics and legal issues are closely related. Ethics takes into account the concerns and values of society as a whole.

Enforcement of the Laws

- Self-regulation
- To achieve true deterrence
- Children's issues
- Microsoft believes that education is the best weapon against piracy

Licensing

Licensing is an increasingly popular method to guard the intellectual property. A license allows the buyer to use the product, but restricts duplication or distribution of any such product in any of its forms. The legal trend favors enforcement of software licenses.

Licenses may be two basic types:

- Shrink wrap or break-the-seal license
- Clickwrap license where the user is required to click a button to accept the terms and conditions of use.

EMERGING ISSUES

- Online governance
  - The Internet Corporation for Assigned Names and Numbers (ICANN) was formed in 1998.
- Jurisdiction is the ability of a court or other authority to gain control over a party.
  - Traditionally based on physical presence.
  - Treaties may provide for international resolution and enforcement.
- Fraud is the use of deception and false claims to obtain profit.
  - The internet provides opportunities for novel deceptions.
  - Spoofing is the use of e-mail or Web sites to impersonate individuals or corporations.
- The FTC, FBI, and state agencies have increased their efforts to track and prosecute fraudulent conduct.

PRIVACY DEBATES AND POLICY

- Privacy supporters advocate policies to inform consumers of data collection and allow them to participate (opt-in) or decline (opt-out).
  - Critics point out that many users do not understand how computers operate and question whether consumers have the expertise necessary to successfully opt-out.
  - Others argue that users wish to receive the benefits of targeted advertising.
  - Access to personal data is another important online privacy issue.
- Although several Congressional bills are pending, no law exists to resolve the privacy debate.

GUIDELINES FOR BUYING SOFTWARE

It is always desirable to choose a product as per ones requirement, so hold true while choosing any software. In order to avoid all such aforementioned legal complicacies or the legal issues involved with the use of software one should always ensure to purchase a software form an authorized vendor of the company and not from any sale outlet. Beside the other common tips involved with the selection and before the purchase of the proprietary software, is that a software should be

- Original equipment manufacturer (OEM) software
- Academic software
- Current usage
- Internal awareness
- Internal IT policies
- License management, software asset management (SAM)
- Software publishers.

CONCLUSION

Software piracy is one of the growing concerns across the world. Computer users from both the developing and the developed world are finding it difficult to cope up with the problem as there is no full proof mechanism in place, wherein a total check can be imposed over the software piracy. Even, the arguments are being made that software developers during the developmental and trial phase of their software distribute these software free among the general public for use and familiarity and the public ultimately end up with getting used to such products, hence develop a sort of dependency on these products. The unregulated use of all such products is being seen increasingly responsible for the growing software piracy.

Ethical considerations are always must always to regulate the use of computers and accessing data on internet via network environment.

Most of the countries across the world have already come up with some IT related laws and regulations, but there is a far greater need to see that general masses be made aware or sensitized about all such laws in place to prevent any misuse of abuse of IT. Also, there is need to frame more deterrent laws whereby misuse and abuse of technology can be prevented. Be sincere in your action, while making use of IT and all other IT -related products and services, otherwise be prepared to face the legal action, which an individual can face if sued in the court of law by the proprietor.

BIBLIOGRAPHY

1. Akdeniz Y, Walker C, Wall D. The Internet, Law and Society. London: Longman Pearson; 2000..

2. Bakos Y. The emerging role of electronic marketplaces on the Internet. Communications of the ACM. 1998;41(8):35-42.

3. Bernstein KJ. The no electronic theft Act: The music industry's new instrument in the fight against Internet piracy. UCLA Ent L Rev. 1999;7:325.

4. Bidgoli H. Handbook of Information Security: Information Warfare, Social, Legal, and International Issues and Security Foundations. Vol. 2 John Wiley & Sons. 2006. p 1008.

5. Bruwelheide JH. The copyright primer for librarians and educators. ERIC. 1995.

6. Cheswick W.R, Bellovin SM, Rubin AD. Firewalls and Internet security: repelling the wily hacker, 2nd edition. Boston: Addison-Wesley Longman Publishing Co., Inc. 2003.

7. Denning, D. E. R. (1999). Information warfare and security (Vol. 4). Reading, MA: Addison-wesley.

8. Himma KE, Tavani HT. The handbook of information and computer ethics. Hoboken, New Jersey: John Wiley & Sons; 2008.

9. Ricketson, S. (1987). Berne Convention for the Protection of Literary and Artistic Works: 1886-1986. Centre for Commercial Law Studies, Queen Mary College: Kluwer.

10. Kelty CM. Two bits: The cultural significance of free software. USA: Duke University Press; 2008.

11. Krutz RL, Vines RD. Cloud security: A comprehensive guide to secure cloud computing: Wiley Publishing. 2010.

12. Ludlow, P. (1996). High noon on the electronic frontier: conceptual issues in cyberspace. MIT Press.

13. Maiwald E. Network security: A beginner's guide: McGraw-Hill Professional. 2001. p 400.

14. Nimmer D. A riff on fair use in the Digital Millennium Copyright Act. University of Pennsylvania Law Review. 2000;148(3):673-742.

15. Pandita R. Internet: A change agent an overview of internet penetration and growth across the world. IJIDT. 2017;7(2):83.

16. Pfleeger CP, Pfleeger SL. Security in computing: Prentice Hall Professional Technical Reference.

17. Radin, M. J., Rothchild, J., & Silverman, G. (2002). Internet commerce: the emerging legal framework. Foundation Press, Incorporated, The.

18. Reed C. Internet law: text and materials. Cambridge: University Press; 2004.

19. Ricketson S. Berne Convention for the Protection of Literary and Artistic Works: 1886-1986. Centre for Commercial Law Studies, Queen Mary College, Kluwer. 1987;4(37):1033-4.

20. Rosenoer J. CyberLaw: The law of the Internet. Springer Science & Business Media.

21. Scott MD. Internet and Technology Law Desk Reference: Panel Publishers. 2000.

22. Shaw M, Blanning R, Strader T, et al. Handbook on electronic commerce: Springer Science & Business Media. 2012.

23. Stallings W, Brown L. Computer security. Principles and Practice. 2008.

24. Taylor RW, Fritsch EJ, Liederbach J. Digital crime and digital terrorism: Prentice Hall Press. 2014.

25. Viega J, McGraw G. Building Secure Software: How to Avoid Security Problems the Right Way, Portable Documents, Pearson Education. 2001.

Chapter 8

Open Access, E-Documents and Information Pollution

INTRODUCTION

Open access sources of information have become almost the order of the day. There is hardly any teaching or research field in which open access sources of information may not have made their presence. In this chapter attempt has been made to draw awareness among students, scholars and teachers about the open access source of information, electronic documents and the prevailing information pollution. Details have been shared about the various open access sources of information, which one can easily access and make use of them for free of cost (be they books, Journals, magazines, databases or even free libraries). Some tips have been given, as how to access these databases and other open access sources of information. The chapter also deals with the aspects like the need and importance of publishing the research results in open access journals, which generally do not charge any kind of publication fee and those who do charge some nominal fee. Some inputs have been given about the existing parameters like, impact factor and H-index, which are being used among the scholarly community to judge the quality of a research publication and how one can also make use of such parameters in choosing the quality journals to publish their research results. Discussions have also been made about the existing information pollution and how the internet has become one of the major concerns of information pollution and the necessary precautions thereof information seekers should take to ensure that the information retrieved is free from contamination.

The following points are discussed in the chapter

- Open access movement
- Impact factor
- H-Index
- E-documents
- Information pollution

WHAT IS OPEN ACCESS?

Open access (OA) means availability of scholarly information and other research results in the form of research publication or other general articles on the Internet, permitting any users to read, download, copy, distribute, print, search, or link to the full texts of these articles, crawl them for indexing, pass them as data to software, or use them for any other lawful purpose, without financial, legal, or technical barriers other than those inseparable from gaining access to the internet itself. The only constraint on reproduction and distribution, and the only role for copyright in this domain, should be to give authors control over the integrity of their work and the right to be properly acknowledged and cited.

TYPES OF OPEN ACCESS DOCUMENTS

There is no end to the type of documents, which are being these days made available to the general masses in open access format. Some of the open access documents, commonly accessed by the students, teachers and scholars of any educational institution include.

- Periodicals (Journals, Magazines, Newspapers, Serials, etc.)
- Books
- Open Libraries

FORMS OF OPEN ACCESS PUBLISHING

Internet has given a new lease of life to the publishing industry. The conventional means of publishing and reading no more restrict the authors from publishing their material and readers from accessing the much needed information on any given topic of their interest.

AUTHOR SELF-POSTING

Interactive web, if on one hand has uplifted the moral of the amateur authors to publish their content, uncensored and uncut on the web by having their own blogs and websites, on the other hand seasoned writers are experiencing a new surge in their literature lover. Web 2.0 has offered new, better and interactive methods of publishing the content and also to immediately receive the reactions of their followers about the published content. This two-way interactive specialty of the web has made it more popular among the internet users. Besides, the greatest advantage of these web postings is that these are accessible in open access format, wherein users need to not pay toward availing these services.

INSTITUTIONAL REPOSITORIES

Universities and other leading research and academic institutions across the world have started offering the access to their research output to the information seeker free of cost. The research institutions are creating web portals of their own and hosting their research data on their respective web servers for free access to information seeker. This service apart from fulfilling the purpose of letting the research results known to the end users also reflects the research strength of any institution. Some of the commonly used software by the institutes to host their research output include:

- DSpace
- Greenstone
- Eprints etc.

Following are the two most popular databases of institutional repositories, wherein people can access the institutional repositories of various research and academic institutions of the world.

- Open DOAR (http://www.opendoar.org/)
- ROAR (http://roar.eprints.org/)

OPEN ACCESS JOURNALS

Open access journals (OAJs) are those journals, which make their published content freely available to the readers through internet. Open access publishing has more or less become the order of the day, as more and more publishers are turning to open access. The basic philosophy behind the open access publishing is that it fulfills every such purpose for which any research activity is undertaken. Each research activity is aimed to serve the purpose of end users or the ultimate beneficiaries for whom the research is generally undertaken. Open access has become one such vibrant medium, which delivers things right at the doorsteps of the end users without paying anything for it. Open access publishing ensures wider and deeper reach of research results with far greater visibility than the subscribed or closed access journals.

- **Born OAJ:** Journals, which are being published in the digital or electronic form since their inception, are known as born open access journals.
- **Converted to OAJ:** Most of the journals until very recent past used to be published only in the printed form, but given the advantages of open access publishing, most of these publishers have started publishing their journals in the electronic or digital form. Publishing again a research journal in the digital form does not mean that the journal is available in the open access form. Most of the journals published in the digital form are still being accessed by the users on the subscription basis. However, a growing trend has evolved among all such conventional publishers to make their published content available to the seeker of information freely, viz. in open access format. All such journals, which have changed their form of publishing from closed access to open access, are termed as converted open access journals.

Most of the research organizations all across the world have realized the need and importance of the open access publishing. Accordingly, a sizeable number of journals are nowadays being published in the electronic format and are being made freely available to the information seeker all across the world. Some of the key databases and data sources, where from information seeker can access quality open access research journals include:

- Directory of Open Access Journals (DoAJ)(http://doaj.org/)
- Open Journal Access System (OJAS) (http://www.inflibnet.ac.in/ojs/)
- Public Library of Science (PLoS) (http://www.plos.org/)
- arXiv.org (Repository of Pre-prints) (http://arxiv.org/)
- Indian Academy of Science (http://www.ias.ac.in/)
- NISCAIR (http://www.niscair.res.in/)
- JURN Directory (http://www.jurn.org/directory/)
- BioMed Central (http://www.biomedcentral.com/journals

OPEN E-BOOKS

Like e-journals and open e-journals, internet is flooded with e-books and open e-books. Open e-books are simply available in the electronic form, which one can easily download freely from the internet. Most of the open e-books accessible on the internet constitute of the classics and various other popular writings. Besides, lots of initiatives have been taken by various agencies across the globe, wherein they have helped to federate all such e-books at a single place.

Some of the common sources for free e-books, where from people can easily access and download books of their interest include:

- Project Gutenberg (http://www.gutenberg.org/)
- Free e-books library (http://www.ebooksgo.org/)

- UN Press E-books collection (http://publishing.cdlib.org/ucpressebooks/search?style=eschol;smode=advanced)
- Open library (https://openlibrary.org/)
- Baen free library (http://www.baenebooks.com/c-1-free-library.aspx)
- Free Tech books (http://www.freetechbooks.com/)
- Intech books (http://www.intechopen.com/)
- Islamic books (http://www.sultan.org/)

Apart from above, the book lovers can also access millions of books on the following open e-libraries.

- The free library (http://www.thefreelibrary.com/)
- The online literature library (http://literature.org/authors/)
- Urdu dost (http://www.urdudost-online-library.com/index.php)
- The electronic library of mathematics
- Questia Free Books (https://www.questia.com/library/free-books)
- Social Science Cyber Library (http://www.socsccybraryamu.ac.in/)

STATISTICAL SOURCES

Statistical sources as the name itself signifies, consists of those sources which deal with the collection, compilation and provision of statistical information in any given field. Most of the government departments are these days making available their information about the progress or regress of their department in the statistical form. UN documents, World Bank documents and various other global agencies reflect their developmental initiatives in the statistical form and so do the various governments.

Some of the key features of the statistical sources of information are:

- The data reflected is empirical in nature
- Reflects trends over time and not the results
- Helps one to draw comparisons and formulate the way-out
- Data is supported with graphics, charts, tables, etc.
- Data is generally presented in a very structured form
- Data provided or accessed can be of any agency, irrespective of the nature, size and the type of organization.

DATABASES

Database is a collective term often used in the field of academics to refer to those sources of information, which are made available in the form of a package to the universities, research institutions and other institutions of higher learning.

Some of the commonly available databases subscribed by the universities and other academic institutions across India are:

- Springer
- Wiley
- Taylor and Francis
- Emerald

- Oxford University Press
- Elsevier
- JSTORE etc.

DATABANKS

Most of the time databanks and databases are used interchangeably by the users. The nature and the type of services people are able to get from both the databases and the databanks are alike. However, there is a significant difference between the two. A databank can be referred to as a data source, which consists of various databases. Most of the academic and research institutions subscribe more than one database and this number may go anything up to or even beyond 100 odd databases. Given the fact it becomes always difficult for the institutions to take up each individual case with each individual database service provider, as such databanks have come as a handy tool to overcome this problem. Academic or research institutions can negotiate subscription for any number of databases with a particular data bank. These databanks apart from making the subscription smooth and easy, which is more like a single window system of dealing, also helps an institution to make some savings by working on discounts, etc. Apart from these, there are various other advantages of databanks, which normally institutions encounter while subscribing different databases from different subscribers.

Some of the common and simple examples of databanks are:

- e-Shodhsindhu
- EBSCO information services etc.

ADVANTAGES OF OPEN ACCESS OVER SUBSCRIBED ACCESS

Although, each type and form of publishing has its own advantages and disadvantages, and thereon, if different forms of publishing exist, it means that there are takers of each individual form of publishing in the market. Still more, the OA documents have some common advantages over that of subscribed ones, which as a result result are more preferred by the user community.

Some of the common advantages of OA publishing over closed access publishing are:

- OA documents are more visible
- Accessibility
- Freely available
- Serves the purpose of research undertaken to a greater extent
- Wider impact
- Serves purpose of a much wider user base at reduced cost.

CREATIVE COMMONS

Creative commons (CC) is a legal global licensing agreement, wherein people can access, use and share their own or other creative work free of cost on the internet. CC licensing agreement also allows a person to build upon the others work under the legal arrangement, however, the person has to pay due attributions to he person on whose work one may build his/her new work.

Some of the common licensing agreements, under which people access, use and share the information, are as under.

- Attribution CC BY
- This license lets others distribute, remix, tweak, and build upon your work, even commercially, as long as they credit you for the original creation.
- Attribution-Share Alike CC BY-SA
- Attribution-No Derivs **CC BY-ND**
- Attribution-Non Commercial CC BY-NC
- Attribution-Non Commercial-Share Alike CC BY-NC-SA
- Attribution-Non Commercial-No Derivs **CC BY-NC-ND**

The licensing agreement enable the people to easily move from all rights reserved to some rights reserved.

SHERPA MODEL OPEN ACCESS INSTITUTIONAL REPOSITORY

The technological boom has given a new fillip to the educational and various other institutions, wherein they created their own digital libraries in the form of institutional repositories. Most of the academic or research institutions host their research output on their institutional repositories for free and complete access of their institutional users only on an intranet. Still more, there are institutions, which provide access to their institutional research output to the outside users on the internet. Providing access to users of institutional repositories is governed by following preprint and post-print publishing rules. The basic principle of preprints and post-prints is based on the color scheme, which signifies the type and kind of documents which people can host on their institutional web servers for the public access.

Preprints and Post-prints

- RoMEO (Rights Metadata for open archiving) color archiving policy
- <u>Green</u> can archive pre-print *and* post-print or publisher's version/PDF
- <u>Blue</u> can archive post-print (i.e. final draft post-refereeing) or publisher's version/PDF
- <u>Yellow</u> can archive pre-print (ie pre-refereeing)
- <u>White</u> archiving not formally supported

It is always imperative for institutions to observe these provisions in letter and spirit, as the violation of same can lead to legal confrontation, which can be challenged in the court of law for violation of copyrights.

IMPACT FACTOR

Impact factor can be viewed as a research evaluation technique, whereby we are able to see and assess the response received by a particular research article among the peer research community. Gross and Gross (1927) gave the concept of reference counting for ranking. Garfield (1955) was the first to count citations to assess the impact of scientific journals. Garfield devised the formula, which is still being used in the same way to compute the impact factor of the journals. the concept of Impact Factor is no more limited to calculating the impact factor of research journals only, as, on date, researchers calculate their individual impact factor, the impact factor is being calculated for institutions, countries, regions, continents, persons to even individual articles, that too for different periods of times, which generally ranges from anything between 2 and 5 years. But, the basic method of computing, the impact has not changed.

The impact factor is calculated for any given period of time, but to evaluate the current impact of publications, Garfield suggested computing impact factor by undertaking the publications of the preceding 2 years for which the impact factor is to be calculated.

Higher the impact factor of a journal, the greater is the rejection rate of articles. It has been observed that most of the researchers across the world try to publish their research results in journals having a higher impact factor. Accordingly, it has been observed that the journals which have a high impact factor, enjoy the greater name in the publishing industry, resulting in rejection of the higher percentage of articles. So, most of the articles published in all such high impact journals are being rendered of high quality among the academic circles and these articles in turn receive a higher number of citations. The Impact factor of a journal is also used as a parameter to submit the article in a reputed and relevant research journal, which also makes these journals to enjoy a far greater readership.

How to calculate the impact factor of a journal, as propounded by the Garfield

Suppose, we are to compute the impact factor for Journal X for the year 2014, simply one has to count the number of articles published by journal X during the year 2012 and 2013 and see how many citations these articles have received in the year 2014. Let x be the number of articles published during 2012 and y be the number of articles published by the journal in the year 2013. Let A be the number of citations received by these articles during 2014.

The impact factor will be x + y/A= I.F

HIRSCH (h) INDEX

Hirsch (2005), a physicist from the University of California gave the concept of the h-Index. H-index is a sort of quality parameter used to assess the citations received by a research article in any given period of time. The concept works exactly on the similar lines to that of impact factor, which can be calculated for a single article, journal, institution, individual, country, continent, subject field, etc. In order to calculate the h-index in any given field, it is the number of citations received by research articles published by a journal, individual, institution, etc., are counted and calculated as per the below mentioned method, proposed by Hirsch.

Hirsch, gave the following method to calculate the h-index.

- A scientist has index h, if h of his/her N_p papers have at least h citations each, and the other $(N_p - h)$ papers have no more than h citations each. *(Jorge E. Hirsch, 2005)*

New Publishing Models

One may come across people, who may be having a strong liking for e-books and other e-documents. At one hand there are people who prefer digital documents over the conventional or the print document, on the other hand, there are people who still show strong like for print documents and drag more pleasure while reading the print. Still more, a good lot of people like to go for either form of published matter, with the result most of the publishers are making their published content available both in the print and the electronic format.

- Electronic journals
- Hybrid: Paper and Electronic (P-E)
- Author self-posting
- Institutional repositories
- Blogs
- Social media

E-DOCUMENTS

E-documents or electronic documents are those sources of information, which are produced, sold, accessed, shared in electronic form by using electronic gadgets like, computers, laptops, mobile phones, etc. A growing trend has evolved across the academia toward the use of electronic documents. Most of the time, people remain absorbed to electronic gadgets, be it at their workplaces or while being at home. Most of the activities, which previously a person used to undertake by moving from one place to another place can these days be undertaken with the help of Information Technology while sitting at home and so has this IT become a reason for change among the book lovers, who prefer to read e-books over the print books. As such, people find it easy to do all their routine activities in electronic form. Electronic documents also refer to those documents as well, which are used to make the electronic gadgets operational.

E-documents or electronic documents are referred to:

- E- Documents are those documents, which require an electronic gadget like computers, mobile phones, etc. to access, display and read the content
- An E-document can be a text file, graphic file, multimedia file, audio-visual file, spreadsheets, etc.
- E-document are generally stored in devices like magnetic disks (floppies, tapes), optic disks (CD, DVD) etc. External hard disks, pen drives, remote servers, on clouds, etc.

Various E-Documents

There are various forms of electronic documents, which are commonly accessed by the academic and research institutions all across the globe. Some of the common forms of e-documents generally accessed by the people are

- Online
- Offline
- Network-based resources
- Consortia based resources
- E-books, e-journals, e-papers
- Radio, TV, Audio cassettes, Films
- Microfilms, Microfiche
- Databases, statistical sources,
- Databanks
- Reference databases (Dictionaries, Encyclopedias, Directories, etc.)

Need and Importance

Importance of E-documents can be assessed from the fact that it is said by the time the information is published it becomes obsolete. The fact also remains that in countries like India print is still the preferred source of information.

Impediments of E-Documents

Over a period of time, a growing trend has evolved among the internet users toward the use of the e-documents. Still more, it has been observed that people do not enjoy access to these e-documents in a uniform pattern. A considerable difference is being observed in the access, use and availability of e-documents across the globe, mostly depending upon the economic and other infrastructural soundness of each individual country. Even the individual economic soundness of a person do makes a considerable difference in accessing the e-resources.

Accordingly, there are numerous impediments which hamper the easy access to e-documents, some of the common impediments faced by the people in accessing the e-documents are as under:

- Lack of infrastructure (labs, Hardware, Software, etc.)
- Lack of user-friendly environment
- Lack of proper training (users and Staff)
- Digital divide
- Lack of IT Skills
- Technophobia

Advantages of E-Documents

Advantages of e-documents are always sought by drawing parallels with the printed documents or the p-documents. Some of the common advantages which these electronic documents have over the conventional documents, include:

- Space saving
- Easy storage
- Easy to use
- Easy updating
- Powerful searching
- Free from theft, binding and repairing issues
- Multimedia features
- Single copy can serve the purpose of many
- Round the clock availability

Instant Messaging (IM) (A substitute for emails): Instant messaging has become a good substitute for conventional electronic mail delivery system. Previously, people used to send the mails, which used to land in the inbox of the recipient and were read by the recipient the moment he/she opens the email account. However, IM has turned out to be a good substitute for the conventional electronic mail, as people are able to send instant message the moment person is online and the person can seek a reply instantly, which has overcome the problem of losing the important mails in the inbox of the recipient and has overcome the delay caused between the delivery and reading of mail.

Disadvantages of E-Documents

E-documents have some common disadvantages, which otherwise are missing in the printed documents. Some of the main disadvantages involved with the use of e-documents include:

- Difficult to read continuously from any electronic gadget
- Develops strain upon the eyes and mind of a reader.
- Need to carry always a gadget to read any document
- Wear and tear of electronic gadgets
- Familiarity with use of electronic gadgets efficiently
- Cumbersome to use online sources of information,
  - Pages are either lengthy,
  - The size of the font is not mostly of advisable size,
  - The color combinations are not suitable for the eyes.
  - Besides online usage always distracts from the main source

Some Other Disadvantages

- Technology is changing with greater peace than human handling
- Need to constantly upgrade with technology
- One form of creation and preservation of documents becomes obsolete with another form of technology
- Electronic dooms day
- Hybrid publishing

SOURCES OF INFORMATION

Some of the common sources of information are shown in Table 1.

Table 1: Sources of information

Conventional	Nonconventional	Online	Offline
Books	Cartographic Material	WWW	Cassettes
Journals	Reports	Emails	CDs
Newspapers	Patients	Online	Multimedia
Magazines	Standards	Databases	
	Dissertations	Blogs	Floppies
	Thesis	Networks	DVDs
	Photographs		E-Books
	Drawings		V-Books
	Graphic material		Optical disks
	Illustrations		Pen drive
			Ex HDD

INFORMATION POLLUTION

Information pollution in the simplest terms can be defined as the prevailing information environment, wherein people are supplied with contaminated information, which is of less importance, irrelevant, unreliable and unauthentic, which lacks exactness and precision, which always has an adverse effect on society at large.

Sources of Information Pollution

One can emphatically say that there are as many sources of information pollution, as many there are sources of information production. Apart from that, ways and means of information handling do become the reasons, which contaminate the information. Organization of information or knowledge on the scientific lines is as important as the creation of information itself. Unorganized information is nothing more than a heap of straw, which instead of becoming the source to seek the solution of the problem, turns out to be the source of distraction and wastage of time. It is believed that one can seek the solution of a problem in lesser time by experimenting it in a laboratory, than looking for a solution in the unorganized information.

Some of the common sources, which are being seen as the sources of information pollution in the present day information world, include,

- Sources of information production

- Publishing industry
- Information technology (Internet/www)
- Information explosion
- Information overload
- Information crisis
- Unorganized information
- Internet clutter
- Plagiarism

Printed Sources

Conventional sources of publishing, by and large are rated as free from information pollution, which otherwise has emerged with the contemporary means of publishing industry. But, still there are some practices involved with the publishing of printed sources of information, which can be seen as contributing rather than leading to information pollution.

Some common trends or practices followed in the conventional publishing, which leads to information pollution, include:

- Numerous journals and books published around the same concept
- Common and confusing titles
- Abundance interpretations
- Catchy, but deceptive titles
- Information clutter
- Unorganized information

Information Technology

Information technology, if on one hand has become the order of the day, on the other hand it is the same IT industry, which has become one of the major sources of information pollution. IT has become one of the largest means of supplying, unsolicited, superfluous, contaminated, unreliable, and unauthentic information.

- WWW
- Social media
- Spam
- Mobile phones

WWW

When we talk about the world wide web, the first thing which strikes in our mind is the huge pool of information and resources spread across the length and breadth of the globe, which stands duly hosted on the web servers for easy access on the internet. The whole information hosted on these servers does not necessarily mean that each piece of information hosted is correct and reliable. Some of the common issues involved with the information spread across the web are:

- Most of the information on the web is cluttered and unstructured
- Source of unstructured information
- Supplier of unsolicited information
- Unregulated information production and dissemination
- Absence of mechanism to check superfluous and distorted information

Richmond, Evehart and Auer (1998) suggested ten C's to evaluate the web resources

1. Content
2. Credibility
3. Critical thinking
4. Copyright
5. Citation
6. Continuity
7. Censorship
8. Connectivity
9. Comparability and
10. Context.

Other Considerations

While retrieving information from the web, it becomes imperative to ascertain the following few facts about the piece of information retrieved.

- Reliability
- Accuracy
- Authority
- Currency
- Coverage
- Objectivity
- Author competence
- Document validity
- Overt and covert affiliation with an institution.

Social Media

One of the greatest offerings of the internet to the global masses is the social media. The social media apart from maintaining the social network of the people, have also helped to overcome the barriers of space and time. Social media is being seen as one of the greatest blessing of the IT. Social media apart from handling the social assignments of the public has also become a forceful public mouthpiece. People are not only able to access and share information through social media, but also have helped to give vent to their ideas, which they share freely among their own people. Apart from these advantages, the social media has somewhere become a reason for concern as well. Of late, people have started misusing this social media for their individual interests as per their convenience and suitability.

Some of the common reasons, as why this social media have become an area of concern:

- Ushered a new dawn of freedom of expression
- A public mouthpiece
- Dangers of absolute freedom of expression
  - A source of uncensored and unsolicited information
  - Inciting violence
  - Derogatory
  - Affecting public opinion
  - Fomenting communal hatred

- Elbert Hubbard, has said, "responsibility is the price of freedom", rightly for the fact that excessive freedom of expression involves dangers of its own kind.

Spam

Spam is an undesirable mail received by the recipient from various known and unknown sources in the electronic format in his/her email account. The main problems associated with the spam is that information received is firstly unsolicited in nature, secondly, most of these mails are either promotional in nature or simply trash and superfluous, serving no real purpose. It is a common practice among internet users to check their morning, evening mails on a daily basis, thereon, if the mails received are more a source of distraction and more a wastage of time, hence a source of information pollution.

- Delivery of undesired stuff
- Least useful information
- Time consuming
- Junk mail
- A personal assault.

Mobile Phones and Information Pollution

Mobile phones have equally become a source of information pollution. Modern day mobile phones, although handheld, by no means function lesser than any modern day desktop or laptop. The modern day mobile phone as a walky talky creates inconvenience to people in its own way. There is need to prevent the mobile phones to ring at the below mentioned sensitive places by requesting people to put their mobile phones on civilized mode.

It is always desirable that mobile phone users, especially those in the academic institutions should have this much civic sense to firstly switch there mobile phones in silent mode and secondly avoid using them while in,

- Libraries,
- Classrooms or
- While being in the mid of a meeting, seminar, conference or an important lecture.

Information Explosion

Information explosion, as we know is associated with the unprecedented growth of information. Over a period of time, the publishing industry has expanded manifold, thereon, electronic publishing has supplemented the conventional publishing industry, and with the result we can see almost a disproportionate output in the publishing industry, which is somewhere over and above the amount of information required for consumption. All this has resulted in the information explosion.

E-publishing has given a new fillip to the publishing industry, with the result we can see, that the content which is being made available in the printed format is being equally reproduced in the electronic format. The concept of hybrid publishing has evolved to cater the information requirements of the both the type of readers, and this is also being seen as a trend among the publishers to not to lose the readership of their published content.

Some common problems associated with the information explosion are:

- Exponential growth of information
- Consumers unable to consume all the relevant information produced
- Not necessarily contaminated or polluted

- Substandard production of information
- A quantitative aspect
- Compromises with the quality aspect
- Reproducing information in different formats.

Information Overload

People see the concept of information overload associated with the information explosion. In a way the concept can be seen as interrelated to a great deal, but still there is a fundamental difference, which differentiates the information explosion from information overload. The information explosion is associated with the production of information, while information overload is associated with consumption of information.

Information overload has left the people with a huge pile of information for consumption, which in fact the information seekers are supposed to consume, but is turning to be over and above the processing capacity of an individual. Information overload can be associated more with the availability of a wide range of information on any given concept.

- Over and above the processing capacity
- Aimed to cater information requirements of both electronic and print readers
- Hybrid publishing
- E-Publishing alone
- Social networking sites, like, Facebook, Twitter, Blogs, Flicker, Digg, LinkedIn, etc.
- The information is unorganized, unauthentic, cluttered, resulting in information pollution (IP)

Information Crisis

Information crisis is seen under two different situations, one is, when we are in need of information on any given topic and the same is not available for consumption, the situation leads to the information crisis. On the other hand, there are occasions when there is over and above information available on any given topic, but with diverse inferences and interpretations, leads to the state of information crisis. The former crisis is due to deficiency of information on any given top, while as the later crisis is due to divergent views on any given topic. Both the situations lead to information crisis.

Some other aspects associated with the information crisis.

- Unavailability of desired information
- Deprived of access to desired information
- Information clutters (chaos and confusion)
- Information overload
- Frequent changes in information requirements
- Poor response
- Information uncertainty
- Conflicting information
- Unreliable
- Uncertain information, etc.

Unorganized Information

Production of information is not an end in itself, as this should be superseded by the organization of knowledge and information. The organization of information become imperative for the fact that, a deluge of information production is being already experienced by the information seekers, which has made it difficult for them to seek the relevant information without wasting time. Time is one such great constraint, which greatly influences the organization of knowledge and information on scientific lines, so that retrieval of information may not become an issue. On the other hand, if there is a large scale production of information only without its proper organization, will surely lead to chaos and confusion, which will ultimately result into information pollution. In order to avoid all such undesirable confusions, there is always need to create order out of disorder.

- Organization of information is as important as the creation of information itself.
- Improper and unorganized information is nothing more than a heap of straw.
- Internet
  - Time consuming
  - Eclipsing the space and time distinction.
  - Source of distraction
  - Too much information available (information overload)

Plagiarism and Information Pollution (IP)

Plagiarism is an information pollution of its own kind. Though, people may dub it only to steeling of others work or intellectual theft without acknowledging the others work, but the fact is reproducing others information by labeling it own leads to confusion to information seeker as what to rely on and the authoritative source of information thereof.

One can easily draw similarities between the environmental pollution and the information pollution and consequences, which both the type of pollutions bore, are almost alike. People are well aware of the consequences of the environmental pollution and the ill effects it has over the life on the earth. Accordingly, consumption of polluted information can lead to such undesirable changes in the society, the repercussions of which may be faced by the generations to come.

Information pollution leads to subscription and consumption of

- Contaminated information
- Unreliable information
- Unauthentic information
- Irrelevant and unsolicited information

Concerns of Information Pollution

Consequences of information pollution can be dubbed even of much higher degree than that of environmental pollution, for the fact that if the polluted information is supplemented to any decision-making process is ultimately going to influence the outcome of that particular activity. Some common areas, which get immensely affected by supply and use of contaminated information include:

- Decision, making
- Health
- Society
- Business
- Education
- And what not

171

Preventive Measures

As is said, prevention is always better than cure, so holds true about adopting the preventive measures, while consuming the information. People have to be very much cautious and careful while selecting and consuming a particular piece of information and this is something which can be achieved only by sensitizing people about the ill implications of information pollution and the consumption of contaminated information. It is equally desirable that preventive measures should be adopted, while producing the information as every care should be taken that each piece of information, which is about to emanate should be free from deficiencies.

Some of the common practices and programs, which can be adopted, whereby people can be sensitized toward the use of uncontaminated information and the other legalities involved with the use of information.

- Information literacy
- Information ethics
- Censorship
- Legal provisions
- Abiding and promoting by social norms
- Abiding and promoting self-regulation while posting content on the web

INFORMATION ETHICS

There is hardly any profession or practice in the world, which is not bound by the ethical consideration. Professional ethics, what we generally know, actually confine the limits within which a professional is supposed to operate, and the violation of same is bound to question the integrity of the person or professional toward his/her profession. Ethics also, somewhere regulate the professional behavior of a person and ensure his/her allegiance toward his/her work. Accordingly, information ethics are there to regulate the behavior of information professionals toward the fair use of information. Information technology has become one of the greatest concerns, which has taken a toll on the ethical consideration of information and information professionals. People are making use of information and IT in such a manner that somewhere they have forgotten their social obligation and the ethical consideration of their profession.

The more worrying aspect associated with the ethical use of information is about those people who violate these ethics like anything and also in turn become the victims of information deceit, without actually having any knowledge or idea about the ethical consideration involved with the use and sharing of information. Information ethics help a great deal in overcoming the issues related to the information pollution and its unregulated distribution. These are as follows:

- In information creation, collection, dissemination and exploitation
- Floridi (2002) advocated of ethics to be extended from the biosphere to the info-sphere
- Ethical use of ICT for sustainable development
- Bridging digital divide
- Fair use of information
- Handling and use of information technology
- Avoiding plagiarism,
- Copyright, intellectual property rights and censorship,
- Information hacking, to various other information abuses.

CENSORSHIP

Censorship is an age old practice to regulate the flow of the desired piece of information in the society. If any piece of information at any point of time is being seen as undesirable, which may have ill effects on the society is generally censored by stopping its circulation among masses. Censorship in the present day world is equally effective and is being used as one of the means wherein an undesired piece of information is kept at bay by following measures.

- Regulating the behavior of people involved with publishing provocative and anti-social content.
- Constructive in the dissemination of authentic and reliable information
- Contrary to above, the absence of censorship of content published on various websites, blogs and social networking sites goes irrepressible
- Creating legal provisions.

CONCLUSION

Electronic documents and open access are the two interdependent concepts. The growing trend toward the publishing of electronic documents has given impetus to the concept of open access publishing. Open access means, the provision of information free to its seekers. On the other hand easy access, use, timely access to content is the strengths, which have given electronic documents a greater degree of acceptance. Subscription of electronic documents means no need to develop physical collection in the library, no need to circulate documents like the print matter and above all any number of users can access the documents at the same time, from anywhere at any point of time. Although, there are equally some growing concerns toward the use, access and subscription, but the advantages of e-documents overshadow its disadvantages. The biggest disadvantage which is being associated with the kind of information pollution created all across the web. The web is pollution in the form of unregulated and uncontrolled growth, the authenticity and reliability of information available across the web, the unsolicited information, the dangers of social media, the internet addiction, etc., are some of the areas, which the users term as an ugly face of the internet.

BIBLIOGRAPHY

1. Antelman K. Do open-access articles have a greater research impact?. College & research libraries. 2004;65(5), 372-82.

2. Bryne JC, Valen E, Tang MH, et al. JASPAR, the open access database of transcription factor-binding profiles: new content and tools in the 2008 update. Nucleic Acids Research. 2008;36(suppl 1):D102-6.

3. Cameron GT, Curtin PA. Tracing sources of information pollution: A survey and experimental test of print media's labeling policy for feature advertising. Journalism & Mass Communication Quarterly. 1995;72(1):178-89.

4. Cameron GT, Ju-Pak KH. Information pollution? Labelling and format of advertorials. Newspaper Research Journal. 2000;21(1):65.

5. Combs J, A. U.S. Patent No. 6,138,129. Washington, DC: U.S. Patent and Trademark Office 2000.

6. Creative commons. About The Licenses What our licenses do. [online] Available from https://creativecommons.org/licenses/ [Accessed December 18, 2017].

7. DeRose S, Vogel J. U.S. Patent No. 5,557,722. Washington, DC: U.S. Patent and Trademark Office 1996.

8. Dillon A. Designing usable electronic text: Ergonomic aspects of human information usage. CRC Press. 2004.

9. Floridi, L. Information ethics: an environmental approach to the digital divide. Philosophy in the Contemporary World. 2002;9(11):39–45.

10. Gadd E, Oppenheim C, Probets S. RoMEO studies 6: rights metadata for open archiving. Program. 2004;38(1):5-14.

11. Garfield E. Citation indexes for science. A new dimension in documentation through association of ideas. International Journal of Epidemiology. 2006;35(5): 1123–1127.

12. Gross PL, Gross EM. College libraries and chemical education. Science. 1927;66:385-9.

13. Harnad S, Brody T. Comparing the impact of open access (OA) vs. non-OA articles in the same journals. D-lib Magazine. 2004;10:(6).

14. Hornbæk K, Frøkjær E. Reading of electronic documents: the usability of linear, fisheye, and overview+ detail interfaces. In Proceedings of the SIGCHI conference on Human factors in computing systems. ACM. 2001. pp. 293-300.

15. Hirsch JE. An index to quantify an individual's scientific research output. Proceedings of the National Academy of Sciences of the United States of America. 2005;102(46):16569-72.

16. Information Pollution, Information pollution. [online] Available from http://en.wikipedia. org/ wiki/ Information_pollution#Related_terms [Accessed December 18, 2017].

17. Lewis M. Plagiarism Controversy: Ambrose Problems Date to PhD Thesis. [online] Available from http://www.forbes. com/2002/05/10/0510ambrose.html [Accessed December 18, 2017].

18. Marcus A, Marcus A. Graphic design for electronic documents and user interfaces (No. 04; Z286. E43, M3.) 1992.

19. Meystre SM, Savova GK, Kipper-Schuler KC, et al. Extracting information from textual documents in the electronic health record: a review of recent research. Yearb Med Inform. 2008:35:128-44.

20. Pandita R. Electronic documents, an integral component of teaching learning process: A critical evaluation. E-library Science Research Journal. 2013;1:31-7.

21. Pandita R. Growing trend towards open access publishing at global level: An analysis of directory of open access journals (DOAJ). International Research: Journal of Library and Information Science, 2013;3(3):2013.

22. Pandita R. Open Access Publishing in India: An Analysis of Directory of Open Access Journals (DOAJ). International Journal of Information Dissemination and Technology. 2013;3(3):176.

23. Pandita R. Information Pollution, a Mounting Threat: Internet a Major Causality. Journal of Information Science Theory and Practice. 2014;2(4):49-60.

24. Pandita R, Ramesha B. Global scenario of open access publishing: a decadal analysis of directory of open access journals (DOAJ). Journal of Information Science Theory and Practice, 2013;1(3):47-59.

25. Richmond B, Evehart N, Auer NJ. CCCCCCC. CCC (Ten Cs) for evaluating Internet resources. Emergency Librarian. 1998;25(5):20-3.

26. Suber P. Open access overview. [online] Available from http://www. earlham. edu/~ peters/fos/overview. htm. [Accessed December 18, 2017].

Chapter 9

Library as a Source of Information

"University is just a group of buildings gathered around a Library."

Shelby Foote

INTRODUCTION

Importance of libraries can be gauged from the fact, if the world faces a doomsday and resourceful libraries like Library of Congress be saved, the world can be easily recreated as it is. This reflects the richness and the essence of a library which perhaps is somewhere the strongest and firmest building blocks of this world, but people undermine the importance of these houses of wisdom because of their ignorance to value such institutions. Gone are the days when people used to find it difficult to locate a particular document in a conventional library system, where collection used to run in lakhs of volumes. Technology has overcome most of such impediments, which hitherto played spoilsport in exploiting the resources and services of a good academic library to its optimum. The library is considered as the heart of any library system and a backbone of any research organization without which neither of these can survive. This chapter deals with the need and importance of a library, its resources and services in the routine research activities of an institution or an organization. A modern day library is a blend of both conventional and contemporary resources and service. While as, the library professionals have been able to exploit technology to its optimum in making good use of library resources and services. The present discussion lasts around some important resources and services available in a modern day academic library, be they open access documents or the subscribed ones. Discussion has also been made about the library consortia, as how these services have proved helpful in accessing a vast number of resources at minimum cost.

Collection Type

Tangible		Nontangible	
Conventional	**Nonconventional**	**Online**	**Offline**
Books	Cartographic material	WWW	Cassettes
Newspapers	Reports	Emails	CDs
Magazines	Patients	Online	Multimedia
	Standards	Databases	
	Dissertations	Blogs	Floppies
	Thesis	Networks	DVDs
	Photographs		E-Books
	Drawings		V-Books
	Graphic material		Optical discs
	Illustrations		Pen drive
			Ex HDD

LAWS OF LIBRARY SCIENCE

Dr S.R Ranganathan, the father of library science in India enunciated the five laws of library sciences in 1931 in his work prolegomena of library science. These five laws of library science form the basis of library science teaching and its practice as a profession.

- Books are for use
- Every reader his/her book
- Every book its reader
- Save the time of the reader
- The library is a growing organism.

These five laws of library science have made an everlasting impact on the library science teaching and the profession and without denial these laws stood relevant to the field and so shall they stand in the times to come.

The first law, as enunciated by the Dr Ranganathan lays emphasis on the very fundamentals of books, is—books are for use. Dr Ranganathan, while highlighting the law in context of library profession talks about involving and adopting all such practices, which will ensure optimum use of books its user. Emphasis has also been laid on, that library professionals should take it as part of their duty, wherein they can maximize the use of books and should not simply be put on stakes for display. In the conventional library system, librarians were mostly known for their custodial nature of duty, however, this notion has changed significantly in the present times, wherein library professionals play a very proactive role in maximizing the use of library documents.

Second law of library science promotes the use of books in its own way. Every reader his/her book depicts about the role of libraries as information and knowledge centers and the role of library professional as intermediaries, whereby they can help to bring into notice of the library users of those library resources, which may be of their interest, keeping in view their query for certain kind of documents. The law focuses on the richness and the resourcefulness of a library, which should fulfill the information requirements of all sorts of library clientele. The second law of library science seeks to bring library users close to the documents of their choice.

Contrary to the second law of library science, the third law of library science discusses about the role of library professionals, wherein they can maximize the use of books by bringing books into the notice of library users. The third law aims to bridge the existing gap between the users and the books. If a library or its professionals fail to seek the readers of its documents, the library can be termed as underutilized.

Time is one of the greatest constraints in almost every sphere of human activity and so gets reflected by the fourth law of library science. There is a considerable difference in the conventional and the contemporary library system, the former system used to purely run by human help and while as the later setup is fully automated, which has minimized the role of human help. In the conventional library setup, it was literally impossible for a library user to locate a document of his/her choice from the huge collection of the library, which used to run in lakhs of documents. Besides, there was no such mechanism in place whereby, it could have been effectively traced whether the document is in the library or not. So most of the time, the whole search of user used to get bogged down by not being able to trace the particular document on a shelf, hence resulting into the wastage of time. The fourth law of library science calls upon library professional to focus on professional ethics, by delivering the services in time bound manner by valuing the time of their clientele. It is always desirable that information be rendered to users in a timely manner, as any delay on the part of information delivery may lead to bog down the whole research project of an information seeker.

Professional skills are one such important area, which needs to be given impetus, as a skillful professional is always in a better position to render services in more efficient, effective and timely manner.

177

Fifth law of library science deliberates over the expansion and growth of the library, which Ranganathan correlated with the living organism. Library shows growth in various spheres, be it in the form of books, clientele, staff, infrastructure or other resources and services. Keeping in view the growth of the library, the fifth law of library science calls for having in place such a library system, whereby every care should be taken to ensure that library grows both horizontally and vertically, be it in terms of accumulation of resources in terms of books and other documents, provision of extended services, growth in readership, library membership, staff, infrastructure, and every other aspect related with the user services. Financial and other constraints should never come in the path of the library, whereby its growth may get hampered. Besides, whatever the growing demands should be met effectively especially in the changing environment of information technology.

WEB RESOURCES

Information and technology has unleashed a new dawn in the information world and therefore it has changed the perception of information professionals toward the role and responsibilities of libraries as institutions and library professionals as intermediaries between the information and the clientele. Accordingly, we can see a change in the approach of library professionals toward their role and responsibility, wherein the conventional library laws still stand intact but in a different form. Alireza Noruzi (2004), in the light of five laws of library science, as enunciated by the Dr S R Ranganathan, propounded these laws in the changing environment of library science and redid them in the growing environment of information technology and the huge pool of e-resources.

- Web resources are for use
- Every user has his or her web resource
- Every web resource its user
- Save the time of the user
- The Web is a growing organism.

Alireza has simply put forth the laws of library science in the light of web resources, which have the same relevance and applicability to the seekers of information as that of conventional sources.

INFORMATION MEDIUM

- Information mediums are for use
- Every patron his information medium
- Every information medium its user
- Save the time of the patron
- The library is a growing organism

(Carol Simpson; 2008)

TYPES OF LIBRARIES

Generally, there are three main types of libraries, these include—academic libraries, public libraries and special libraries, all the three different types of libraries serve the interests of its user base, keeping in view the information interests of their clientele and so do they develop the collection, given the requirements of the users of the library.

Academic Library

Academic libraries are those libraries, which are in academic institutions, be it universities, colleges, secondary or senior secondary schools. The primary aim of all types of academic libraries is to make available all types of reading material to the students, scholars and teaching community which pertains to their academic interest. Textbooks, general books and other reference books constitute the main collection of these libraries.

Public Library

Unlike academic libraries, the public libraries are aimed to serve the interests of the general public. Public libraries are generally established at district level, block level or even at mohalla or village level. Given the nature of public libraries, while developing the collection of these libraries, the main thrust is laid on the acquisition of general and reference books. Public libraries have proved as a good source of leisure for old and retired people, who spend a good part of their time in these libraries. Newspaper and other popular magazines are the most preferred documents, which the users of public library generally prefer to consult.

Special Library

Special libraries, as the name itself signifies are special in nature and their specialty gets signified by the collection these libraries develop over a period of time. Special libraries are generally associated with the research institutions and these libraries develop a collection on a project basis. Special libraries are generally small in size, as these libraries confine their collection on a specific topic or project, and the moment the project is over or completed, the library may stand no more relevant to the next project, as such the collection is discarded by sending it to the parent library and the new collection is developed as per the new research requirements. These libraries are generally created in the close proximity of the researchers and the research labs, so that the information requirements of the researchers may be served without wasting their time.

Of late, the concept of special libraries has undergone transformation and we can see, libraries associated with a specific topic, discipline, subject or program etc. are also being termed as special libraries. Just like a library having a collection of Braille books is termed as special, accordingly, if a library has musical score or notes as collection is termed as special. Similarly, any such library, which possesses any speciality of whatsoever nature is regarded as special in nature.

LIBRARY RESOURCES

It is not the structure of the library what matters, but the resources which a library holds. The richness or the resourcefulness of the library is counted by the number and type of documents it possesses. Gone are the days, when books were the only form and type of resources available with the libraries. A modern day library has to be a collection house of all the diverse forms of information sources, published in whatever form or format. The two main types of resources available with each kind of library are electronic and printed sources of information.

Printed Sources

The major portion of library collection consists of printed matter. In fact, it is the printed matter, which forms the backbone of the library collection. Electronic documents were part of the library collection from the days when they were made available in the form of microfilms and microfiche. But, that did not undermine the importance of one form of documents over the other form and each kind of document has its own readership. Even today, print is the integral part of the every modern day library collection, no matter if there is a growing demand for electronic documents.

Books

When we talk about library, the very first visualization, which strikes to the information seeker, is the books. Books constitute the first and foremost component of the library collection and the library system. The type of book collection, which a library develops, primarily, depends upon the nature and type of the library. However, in an academic library, we generally find the following three types of books

- Textbooks: These are the books which are generally procured, give the course syllabi of a particular subject field. Textbooks, course books, lab manuals etc.
- General books: These are very general in nature, these may cover areas like, poetry, prose, fiction, non-fiction, contemporary writing, historical facts, etc.
- Reference books: Reference books, as we know are those books which are not consulted for continuous reading and are generally consulted to seek the background information of that subject or object, etc. Dictionaries, Encyclopedia's, Almanacs, Cartographic material, etc. fall under its purview.

Periodicals

Periodicals enjoy a special place in the academic libraries. Going by the literal meaning of the periodical, then periodical simply means those publications, which are being published at regular intervals of time. Even though, many a times, people brand those publications are also as periodicals, which are being published regularly, but not at regular intervals of time. Some common forms of periodicals, which are being subscribed in libraries in good numbers include.

- Journals: An academic library, especially university or college library, if it does not subscribe to research journals, is definitely missing out important academic and research supplement for its researchers. Journals are considered as the backbone for any research institution, without which it cannot sustain. Since the journals are published at regular intervals of time and their periodicity mostly varies from monthly to quarterly, half yearly, yearly or even there are journals, which are biennial in nature.
- Magazines: Magazines are of different nature; they can be popular in nature or may be specific to a particular subject. Magazines constitute an important part of the periodical section of an academic library.
- Newspapers: Newspaper are the periodicals which are generally published on a daily basis, however, there are some newspaper, whose periodicity may vary from anything between a week to a year or so.
- Serials: Serials are those periodicals, which are bought out at regular intervals of time, but they deal with the specific subject matter
- Newsletters: Newsletters are regular publications, but are specific to one main topic. A newsletter can be a community-based magazine, a school newsletter, an institutional newsletter, etc. The subscribers of these newsletters generally keep themselves informed about the developments and other happenings in their field.
- Bulletins: Bulletins are generally considered those publications, which are specific to an institution or an organization, mostly publishing information of public interest in the form of official announcements or policy decisions. A newsletter is also sometimes referred as a bulletin.

Electronic Sources

Electronic sources of information have integrated and assimilated into the lives of common masses to such an extent that if taken out at any point of time may lead to the crumbling of the whole information system. A modern day library may find it difficult to survive in the absence of electronic documents. The publishing industry has experienced a great drift, whereby it has become imperative for these publishing houses to adopt the hybrid publishing trend to fulfill the requirements of both the type of information users. A modern day library heavily relies on the electronic sources of information to fulfill the information requirements of their clientele. Some of the common forms of electronic documents are online journals, e-books, web resources, official websites of institutions, governments, organizations, etc. and many more. A good number of publishers make available e-documents in the CD-format as well.

Other Electronic Sources

Some of the common forms of digital documents collected by modern day libraries include:

- Born digital: Born digital are those documents, which are published in the digital form by the publishers itself. These documents are generally made available both in the online and offline form.

- Digitized: Unlike born digital, digitized documents are those documents which are not born in the digital form, but these documents are converted to digital form at a later stage, given the issues related to their popularity or as a means of preserving important manuscripts or other rare material and books are being digitized by pressing into services of high end scanners.

- Online: Online sources of information have gained huge popularity among the academia and are in great demand among students, teachers and scholars. Online sources of information are those sources, which stand hosted on remote web servers, while as users can access them on the internet by logging in on a particular web address. These sources can be easily browsed or downloaded from any remote location, provided one has an authorized access to all such sources. Institutions or organizations interested in subscribing or accessing the online sources of information are required to enter into an MoU with the service provider and has to pay for the sources and services, which an institution or an organization is about to access.

- Offline: Offline source of information, unlike online sources can be procured from the open market like any other document, but can be accessed on electronic gadgets like, computers and other similar devices. Most of the offline sources of information are born digital and are available in the CD-format.

- Open access: Open access documents are online electronic documents, which, unlike other subscribed online sources are freely available on the Internet. One can easily browse or download, share these sources without paying anything for them. Open access sources of information are gaining huge popularity among the researchers, students and teachers. Even a good number of researchers are publishing their research results in the open access journals.

- Subscribed: These are those sources of information, which can be accessed against the payment.

- Databases: Subscription and access to databases is a very common practice among the academic institutions. Under the databases, a huge number of resources are accessed for a wider user base in a single package. Access to journals in the form of databases is a very common practice among academic and research institutions. A database is simply an organized collection of varied information on different subject areas. It is a processed information ready for consumption.

- Data banks: As the word bank itself signifies so as the data bank is a repository associated with the data. Data, as we know is information about any subject or object, it can be specific or diverse. Data banks also refer to those intuitions, which deal with the handling or creating repositories of databases. Each larger organization or big corporate houses have their individual data banks, which deal with the handling of huge amounts of data and its processing almost on a daily basis.

LIBRARY SERVICES

A library of whatsoever nature provides different kind of services to its users. Most of the services provided to the library users also depend upon the nature of the library and the type of clientele, it is aimed to serve. Some of the common types of library services provided to library users include

Circulation Services

Circulation service forms the backbone of any library service. Be it an academic library, a public library or even a special library. Lending documents for its users from the various types of documents collected over a period of time constitutes the foremost service of any library system. Each section of a library has a separate circulation desk, which is mostly maintained at the single door entry-exit point of that very section of any library.

Information Services

It is not necessary that a library clientele may always visit a library for charging and discharging of documents. Libraries associated with reputed institutions or organizations appoint an information officer or information scientists in their libraries with the aim to provide other necessary information to the library users, which may not be necessarily available in the library itself. An information service is a form of help desk for the library users, wherein they are also guided to the various sources, which the seekers of information are not generally able to locate.

Reference Services

Each library has an independent reference section, which is always managed by a full time reference librarian. The reference section of a library holds the reference collection or reference books, which are generally consulted by the users for specific information. The role of a reference librarian is not only limited to the provision of reference books but also about helping the seekers of information to get the desired piece of information, they may be looking for. A reference librarian is supposed to record the reference queries of the seekers and get the same arranged form difference source, so that same be delivered to the seekers. Besides, a reference librarian is supposed to be the person of very wide knowledge, having a deep understanding about different subjects with lots of information, who can satisfy most of information queries of the information seekers from his acquired knowledge.

Referral Services

Referral and reference services are mostly used interchangeable, but there is a significant difference between the two. A referral service unlike the reference service refers an information seeker to very particular source of information, from where one can get the desired piece of information. The source can be an individual, a subject specialist, an institution, an organization or any other documentary or nondocumentary source.

Interlibrary Loan Services

Any library of the world, of whatsoever size and nature is not enough resourceful or rich in its collection to have each and every document in its collection. Therefore, each library cannot satisfy the information needs of its each clientele from its own resources. To overcome this problem, interlibrary loan is a very handy tool and technique, whereby information needs of the users can be met out with greater ease. Libraries of whatsoever nature can enter into an MoU with each other, wherein they can meet out the information requirements of each other by lending books to library clientele of the member libraries on the interlibrary loan basis.

Internet Services

With the growing trends toward the use of electronic documents among students, scholars and teachers, a modern day academic library is somewhat bound to provide round the clock Internet services to its users. A modern day library without Internet lab will not be able to do justice to the services it is providing to its users. Most of the resources subscribed by a modern day library are mostly electronic in nature and to access all such resources, it becomes imperative to have a full fledged Internet lab, where users can access all such resources both online and offline.

Reading Room Facility

Library apart from being a home of huge number of documents and other information sources, do enjoys a distinction of being a place, which is most ideal for reading. Library members are not only interested in using the documents a library possesses in its holding list, also equally keen to make use of that serene atmosphere totally meant for reading purpose. Most of the libraries carve out sufficient space in its each section for reading purpose of its clientele. Even it will not be perhaps inappropriate to say that majority of the library users still look toward libraries as places meant for reading purpose.

Documentation Services

Library documentation services are one of the conventional practices, wherein information professionals were supposed to meet out the information needs of users by extending professional help. The practice of documentation service is still around, but generally this kind of service is rendered to researchers and scientists in the research organizations. In the documentation service it is the library professionals who retrieve the desired piece of information from their own and other resources, keeping in view the information needs of their clientele. Documentation service is generally provided by two means, one is current awareness service (CAS) and another is selective dissemination of information (SDI).

Current awareness service and selective dissemination information:

- CAS and SDI services: Information scientists or information professionals working in research organizations generally collect and disseminate information, keeping in view the information requirements of any given research project. Under the CAS service, the information professional collects all the relevant information on any given topic in anticipation and provide the same to the researchers or scientists, as the same may be of their use. While as, when taken the case of SDI services, under this service, the information professional are given a set of information demands or information requirements by the researchers, which these professional look for and collect the same from different sources, and provide the same to the researchers well in time.

LIBRARY SECTIONS

In order to carry out various technicalities of library documents and to render various services to its users, a library stands divided into different sections, which generally again depends upon the nature, type and size of the library. Some of the common sections an academic library possesses, include the following:

Acquisition Section

Acquisition of documents in the library is one such main activity, which continues round the year. All the different types of documents, which a library procures, are processed by the acquisition section of the library, and it is this very section, which upon arrival of books verifies books with bills and recommends same for the release of payment after completing the accession of the documents.

Circulation Section

A library generally has different sections of documents with each section having a specialty. Each document section of the library has a circulation desk of its own. Circulation of documents is one of the fundamental activities of each kind of library. The first three laws of library science get justified only when a library has an efficient circulation system, wherein users get the documents of their interest easily issued.

Classification Section

Everything we have around us or whatever nature has created is in a proper order. Nature has segregated everything on the basis of its nature and other intrinsic and extrinsic qualities. We can see all like things have been bought together. Accordingly, in the library sciences, when a library develops a huge collection of documents on diverse topics, it becomes imperative to segregate the books on the basis of like and unlike things. Classification section in a library is an area, where the practice of putting like documents together is undertaken. Although, each individual document has a unique call no, but it is class no which identifies a document belonging to a particular lot of documents. Library classification is actually the basic tool in the library science profession, which helps in organizing the knowledge and information on the scientific lines. There are different classification schemes and each individual library has a freedom to choose any of these classification schemes to organize their documents, mostly depending upon the nature and type of a library. Classification of documents in a library is a very important technicality, which more or less is a backbone of the library services and activity.

Cataloguing Section

Cataloguing section, unlike the classification section, in the simplest terms, library catalogue is a list of documents in the possession of a library. Each individual document of a library has an independent catalogue card, depicting most of the bibliographic information of that very document. The earlier card form of catalogue has got replaced by the modern day Online public access catalogue (OPAC)

Periodical Section

This section of the library remains generally thronged by the scholars and other researchers of an institution. The periodical section of the library subscribes all the related research journals, maintains them, routing them displays them and also provides the reading or seating space to the library clientele in their section.

Internet Lab

The Internet lab is a very important and an integral section of any modern day library. It is always impossible for an individual library to meet out all the information requirements of an information seeker. On the other hand, keeping in view the amount of information available on the internet, libraries are somewhat impelled to offer facilities like a computer lab with internet facility. Besides, a modern day library subscribes to an N number of e-resources for its clientele and to have access to these e-resources, there is every need to put in place a modern high-tech internet lab.

Reading Room

Reading space is generally provided in all such sections of the library, in which one or the other type of library collection is developed. The basic concept behind the provision of reading room facilities is to help the users to go through the desired piece of information as and when required. Thereon, over a period of time, the library reading rooms have actually become the most preferred placed for researchers, scholars and other serious readers. Devote library users love to spend most of their time in the library reading room. A modern day library ensures to earmark exclusive space for reading purpose on its each floor. The reading space is generally created in the space where there is provision for ample natural and artificial light with proper ventilation.

Reference and Information Services Section

Reference and Information services, are although independent of each other, but are generally rendered in the reference section of a library. It is the reference librarian, who is supposed to address the information queries of information seekers. A reference section of a library develops a collection of reference documents, which are generally consulted by information seekers for a brief period and not for continuous reading. Information services can be of any nature and the professional assigned with the job of providing information service at the help desk should be having the helpful nature with sufficient knowledge and information to address most of the user queries from the acquired knowledge.

Stack Area Section

Stacking documents on the shelves is one of the main activities involved with the library collection development. In each individual library, a specific area is identified for book stacking, which mostly depends upon the nature and type of the library and the area is generally known as stack area. An academic library associated with the university system always has different sections. All the areas in the library where documents are arranged on stakes are called as stack area section. The documents on the shelves are always arranged in the most helpful sequence so as to facilitate the easy withdrawal of books. The arrangement of books on the shelves is undertaken with the help of class numbers or the call numbers. These class numbers help to arrange books in the stacks in the most helpful and convenient form.

Newspaper and Magazines Section

This is one of those sections, where most of the library clientele visit daily, although for a brief period. Newspaper and magazine sections are generally located adjacent to the entry corridor of the any library building. The most compelling reason to have the newspaper section adjacent to the entry-exit of the library is, it is being observed that the first and foremost thing the library users do in the library is to go through the newspapers and other news magazines to get the latest update about the developments, which may have taken place in and around.

CD Rom/Digital Documents Section

There is no end to the road of digital documents. The digital or electronic documents had been in the possession list of libraries from the very beginning. Previously libraries, along with other printed documents used to have electronic documents in the form of microfilms, microfiche, etc. Over the period of time, these electronic documents got totally replaced by more modern and more attractive e-documents. A modern day library apart from developing an online collection does develop a collection of offline e-documents, mostly in the form of CDs, DVDs, E-books, and various other forms of data storage.

Theses/Dissertation Section

This section reflects the research strength of an institution of which the library is a part. In the theses and dissertations section, an institutional library arranges and displays all such research theses, which have been undertaken by the scholars and other researchers of that institution from time to time. This section generally holds, masters degree dissertations of the pass outs, M Phill and PhD thesis, etc.

Library Technicalities

When we talk about the scientific arrangement of documents in the library, it means we adhere to some basic tools and techniques of library practice. Without undertaking some basic technicalities, it will always be difficult for any library to render efficient library services. Some of the basic technicalities which are followed by almost all of the libraries include

- Classification
- Cataloguing (OPAC)
- Data entry
- Spine labels
- Book plates
- Due date labels

DEWEY DECIMAL CLASSIFICATION

Dewey decimal classification (DDC) is one of the widely used library classification schemes in the libraries across the world. It is believed that more than 2 lakh libraries across 130 countries of the world use DDC in their libraries. The foremost advantage of the DDC is, it is very simple to use while undertaking a classification of library documents and still more simple to follow, while retrieving documents from the library.

It was Melvil Dewey, who in 1876 bought out the first edition of the DDC. The scheme uses Indo-Arabic numerals and in all is divided into 1,000 classes. The online computer library Centre regularly updates the scheme and of late in the year 2011, OCLC released the 23rd edition of the scheme. DDC is an enumerative scheme of classification, which basically works on the three digit notation to the document, which helps it to fix its location among other related documents in a library. Following are the ten main basic classes of the scheme, which get multiplied at ten stages to make is a scheme of 1000 classes.

- 000 – General works, computer science and information
- 100 – Philosophy and psychology
- 200 – Religion
- 300 – Social sciences
- 400 – Language
- 500 – Pure science
- 600 – Technology
- 700 – Arts and recreation
- 800 – Literature
- 900 – History and geography

DDC SUB CLASSES UNDER THE MAIN CLASS 000

(General works, computer science and information)

- 000 Generalities
- 010 Bibliography
- 020 Library and information sciences
- 021 Library relationships
- 030 General encyclopaedic works
- 040 Not assigned or no longer used
- 050 General serials and their indexes

- 060 General organization and museology
- 070 News media, journalism and publishing
- 080 General collections
- 091 Manuscripts

EXTENSION ACTIVITIES

Apart from providing routine services to its clientele, a library also undertakes different types of extension activities for its users. The basic idea of organizing these extension activities is to draw greater awareness among the library users toward making better and wider use of library resources. Library extension activities are also organized to strengthen the bonding between the library and its users. Library extension activities also act as platform, whereby library showcases its prowess in the form of the collection and services to its readers or clientele. The library also promotes itself among a wider cross section of the society as an institution, which holds relevance to them by all worldly means and the doors of it are open for one and all alike. Some of the common extension activities organized by the libraries include.

- Organizing book exhibitions is a regular feature of the institutional libraries in general and university libraries in particular
- Library orientation
- User education.

Other Purposes of Library Extension Activities

- To present library as a social, cultural and intellectual centre
- To inculcate reading habit among non-users
- To bring books and readers together
- To inform those who do not use the library services and to attract them to those services
- To inform the reader of all the facilities offered by the library
- To remind both the readers and the nonreaders of the library about its resources and services
- As a means of publicity to enlist financial support or otherwise for the libraries.

INFLIBNET

The Information and Library Network (INFLIBNET) came into being in the year 1991, but it was only in the year 1996 when it attained the status of Independent, autonomous Inter-University Centre (IUC) of the University Grants Commission (UGC). Automation and networking of university libraries was the primary aim with which the centre was established. To create a union database of the resources available across the libraries of institutions of higher learning in India, so as to promote the resource sharing among academic libraries. Apart from various other services and activities. INFLIBNET is also involved with the subscription of e-journals on a consortium basis for universities and other higher educational institutions recognized by UGC under section 12(B) and 2(F).

E-DOCUMENTS

A modern day library is bound to develop a collection of both electronic and printed documents. If print is what we see and correlate with the conventional library practices, then electronic documents are the ones, which we see as the basis of a contemporary library practices and services. Given the fact, a modern day library has to develop its collection in such a way, whereby it has to cater the information needs of both the type of library users. As on date UGC-Infonet e-resources service is the most popular, as more than 200 universities across the country are availing the services under the scheme.

SUBSCRIPTIONS UNDER UGC INFONET CONSORTIA (E-SHODHSINDHU)

UGC-Infonet Consortia was launched by the INFLIBNET in 2003 by the then President of India, Dr APJ Abdul Kalam. Under the scheme, UGC connected various university libraries of the country by Internet, with the primary aim to provide them access to e-journals subscribed by INFLIBNET under consortia. Under the consortia, as on date, more than 8000 online peer review journals are being subscribed by the centre for its member libraries. Some of the key subject areas covered under the consortia, include, Arts, Humanities, Social Sciences, Physical Sciences, Pure Sciences, etc. The access was provided to the member libraries in phased manner. In 1st and 2nd phase 50 universities were covered, in the 3rd phase, 99 universities were covered along with 14 national law schools and universities and a few other institutes. While as, on date all the universities and colleges recognized by the UGC under section 12(B) and 2(F) are being provided access to e-resources under consortia subscription.

CONCLUSION

Subscription and utilization of e-content has become the order of the modern day academic and research activity. No institution can afford to function elusive of such sources. Keeping in view the growing demand for online sources of information, most of the publishers all across the world have started publishing their content in hybrid form, viz. both online and print. The online publishing by no means has demeaned the demand for print and the importance of the libraries. Libraries are still the most preferred places to access and retrieve the desired piece of information. Although, information and technology has unfolded new vistas of content management and so have these new technological offing's revolutionized the role of library professionals. In the changing environment of IT, if on the one hand information-seeking behavior of information users has changed, so has changed the role of library professionals, who more or less are now being referred as cyberarians.

BIBLIOGRAPHY

1. Auger CP. Information sources in patents. KG Saur Verlag Gmbh & Co 1992.

2. Bawden D. Information and digital literacies: a review of concepts. Journal of Documentation, 2001;57(2):218-259.

3. Buckland MK. Information as thing. Journal of the American Society for Information Science (1986-1998). 1991;42(5):351.

4. Calhoun K. Technology, productivity and change in library technical services. Library Collections, Acquisitions, and Technical Services. 2003;27(3):281–89.

5. Center I N F L I B N E T. INFLIBNET News-Letter Chicago. 1997;3(2–3)

6. Chowdhury G G. Introduction to modern information retrieval. Facet Publishing 2010.

7. Daniells L M. Business information sources. Univ of California Press 1993.

8. Evans G E, Heft S M, Bloomberg M. Introduction to technical services. Libraries Unlimited Inc. 1994.

9. Folster MB. A study of the use of information sources by social science researchers. Journal of Academic Librarianship. 1989;15(1):7-11.

10. Gates JK. Guide to the use of libraries and information sources. McGraw-Hill Companies 1989.

11. Graber DA. Processing the news: How people tame the information tide. University Press of Amer. 1988.

12. Gupta S. E-ShodhSindhu consortium: awareness and use. SRELS Journal of Information Management. 2017;54(2):51–9.

13. Maness JM. Library 2.0 theory: Web 2.0 and its implications for libraries. Webology. 2006;3(2).

14. McCombs GM. Access Services: The Convergence of Reference and Technical Services (No. 34). CRC Press 1992.

15. Mooney RJ, Roy L. Content-based book recommending using learning for text categorization. In Proceedings of the fifth ACM conference on Digital libraries. ACM. 2000 pp. 195-204.

16. Moran BB, Morner CJ. Library and information center management. ABC-CLIO. 2017

17. O'Connor S, Huang J, Wong K. Technical services and user service improvement. Library management 2006;27(6/7):505-14.

18. OCLC Dewey Services Organize your materials with the world's most widely used library classification system. [online] Available from http://www.oclc.org/en/dewey.html [Accessed December 18, 2017].

19. Rubin RE. Foundations of library and information science. American Library Association. 2017

20. Schmidt J. Promoting library services in a Google world. Library Management. 2007;28(6/7):337-46.

21. Soergel D. Organizing information: principles of data base and retrieval systems. Elsevier 1985.

22. Stewart DW, Kamins MA. Secondary research: Information sources and methods Sage 1993.

23. Vickery B. A century of scientific and technical information. Journal of documentation. 1999;55(5):476-527.